Partnering Manual
for Design
and Construction

Other McGraw-Hill Books of Interest

Barkley and Saylor
CUSTOMER-DRIVEN PROJECT MANAGEMENT

Civitello
CONSTRUCTION OPERATIONS MANUAL OF POLICIES & PROCEDURES

Cleland and Gareis
GLOBAL PROJECT MANAGEMENT HANDBOOK

Godfrey
PARTNERING IN DESIGN AND CONSTRUCTION

Leavitt and Nunn
TOTAL QUALITY THROUGH PROJECT MANAGEMENT

Levy
PROJECT MANAGEMENT IN CONSTRUCTION

O'Brien
CPM IN CONSTRUCTION MANAGEMENT

Palmer, Coombs, and Smith
CONSTRUCTION ACCOUNTING & FINANCIAL MANAGEMENT

Ritz
TOTAL CONSTRUCTION PROJECT MANAGEMENT

Ritz
TOTAL ENGINEERING PROJECT MANAGEMENT

Turner
THE HANDBOOK OF PROJECT-BASED MANAGEMENT

Partnering Manual for Design and Construction

William C. Ronco, Ph.D.
Jean S. Ronco, Ed. M.

McGraw-Hill

New York San Francisco Washington, D.C. Auckland Bogotá
Caracas Lisbon London Madrid Mexico City Milan
Montreal New Delhi San Juan Singapore
Sydney Tokyo Toronto

We dedicate this book to our clients:
the organizations and the people in them who have worked with
us insightfully and diligently to clarify and illustrate
the real meaning of effective partnering.

William Ronco
Jean Ronco

Library of Congress Catalog Card Number: 95-46348

McGraw-Hill

A Division of The McGraw·Hill Companies

1 2 3 4 5 6 7 8 9 0   QBP/QBP   9 0 9 8 7 6 5

ISBN 0-07-053669-4

The sponsoring editor for this book was Larry Hager, the editing supervisor was Caroline Levine, and the production supervisor was Pamela Pelton. It was set in Palatino by Cynthia L. Lewis of McGraw-Hill's Professional Book Group composition unit.

Printed and bound by Quebecor-Book Press

McGraw-Hill books are available at special quantity discounts to use as premiums and sales promotions, or for use in corporate training programs. For more information, please write to the Director of Special Sales, McGraw-Hill, 11 West 19th Street, New York, NY 10011. Or contact your local bookstore.

 This book is printed on recycled, acid-free paper containing 10% postconsumer waste.

Contents

3. Specifics   33

4. Benefits: What Partnering Accomplishes   49

5. Deeper Impacts: What Partnering Does and Why   65

6. Problems   87

Part 4. Case Studies   239

14. Partnering by the Book: The Benson Management
Services Building Case   241

15. Partnering as Repair: The State Records
Building Case   275

Appendix   A Construction Lawyer Looks at
Partnering   305

Acknowledgments

We are grateful to the numerous clients and colleagues who shared their thoughts and perceptions about partnering with us and who thus contributed to our own thinking: Ray Bayley at Stein and Company, Steve Einhorn at Einhorn-Yaffee-Prescott, Jim Franklin at the American Institute of Architects, Hank Keating at the Boston Housing Authority, Chris Noble at Hill and Barlow, Dave Oboler at Einhorn-Yaffee-Prescott, Tony Oliver at the National Parks Service, Ken Perrin at the U.S. Postal Service, Dave Raduziner at Sun Microsystems, Gary Schreck at Ameritech Corporate Real Estate, Sheila Sheridan at Harvard University and the International Facility Managers Association, Pat Stoller at A.S.C. Services, and Frank Zuccardy at the State of New York Construction Fund. Steve Hultgren at CRSS Constructors provided detailed, thoughtful, helpful comments on the manuscript.

The American Institute of Architects offers a booklet on partnering (202-636-7300). The Associated General Contractors offer both a booklet and a video (202-393-2040). We found all these to be very useful in mapping out a clear, general approach to partnering. Richard Fitzgerald at the Boston Society of Architects and Margaret Neil at the Boston chapter of the Associated General Contractors provided additional personal insight into effective partnering.

It was very good to work with Bruce Sargent at Lotus Development on the groupware chapter. He and Lotus's consultant to the project, Robert Larsen-Hughes, responded not just patiently but with continuing enthusiasm to the many questions we posed about the specifics of implementing groupware.

The Takenaka Corporation provided funding support for research conducted by the Space and Organization Research Group (SPORG) at the Massachusetts Institute of Technology. SPORG provided funding for our writing of case studies on partnering and groupware. These case studies provided the basis for Chapter 13 on using groupware and the case study in Chapter 14 of Benson Management Services. In addition, the SPORG discussion group provided, in a brief period of time, useful insight into the impacts of partnering on the design process. This group included Turid Horgan, Michael Joroff, William Porter, Donald Schon and Sheila Sheridan. Michael Joroff managed SPORG's interaction with my project and contributed many useful insights.

We want to thank the contractors, subcontractors, architects, engineers, planners, and clients who have participated in our partnering workshops. We recognize that while it is potentially very rewarding, partnering also poses new challenges and places new demands on people who have already an abundance of responsibilities and obligations. We have appreciated the open mindedness and energy which workshop participants have devoted to making partnering effective.

Finally, we appreciate the support provided by Larry Hager, our sponsoring editor, and Caroline Levine, our editing supervisor.

William C. Ronco
Jean S. Ronco

Partnering Manual
for Design
and Construction

Introduction

Improving Partnering Effectiveness

Improving Partnering Effectiveness

Partnering is an organized effort to improve communications in design and construction projects. If designed and implemented well, it can improve project communications, profitability, and quality while reducing costs, conflict, and exposure to litigation.

We have formed several opinions about partnering that frame this book:

- *Partnering can be very effective.* We know partnering is capable of attaining significant results because we have had the opportunity, first hand, to design and facilitate numerous partnering efforts. We have worked with architects, engineers, clients, and contractors and shared in their struggles and exhilaration as they devised more productive ways to work together.

- *Partnering can influence not just the surface but the very core of the design and construction industry's standard business practices.* More than a workshop or a "program," partnering provides strategies, skills, and connections that have the potential to redraw the long-standing, fragmented boundaries of the industry.

- *All partnering efforts are not alike.* There is widespread agreement about some of the topics on a typical partnering workshop agenda but also widespread diversity in program content and focus.

- *Some partnering efforts are more effective than others.* It is not enough to hang a partnering poster in a construction trailer or hold a partnering workshop in a hotel to generate results.

Partnering stands at a crossroads in its evolution; it is becoming important to define what makes it effective. Key professional societies (Associated General Contractors, American Institute Of Architects, American Council Of Consulting Engineers, etc.) endorse it. Many federal, state, and local agencies require it for large contracts. Corporations are using it in their corporate real estate processes.

Yet partnering has not so much evolved that it is absolutely consistent. Participants in partnering programs find some similarities from one to another, but they usually also find some significant differences. Some of the professional societies that endorse partnering add to the diversity (and sometimes the confusion) of partnering practices by advocating methods locked within the paradigms of their respective constituencies.

As partnering becomes an increasingly common business practice in the design and construction industry, we want to influence positively the shape it takes. We cannot claim to possess complete knowledge of partnering or to hold the positional understanding of any one of the building professions or trades. However, we do bring to the task of defining the nature of effective partnering unique understanding, perspective, and objectivity.

We have not only consulted for numerous partnering efforts, we also have provided team-building consultation and management training to organizations involved in all aspects of the industry: architecture and engineering firms, corporate real estate organizations, public agencies, contractors and subcontractors, and crafts and tradespeople.

We know the biases and strengths of each profession, trade, and organization. We know their successes and frustrations, their internal communications problems, and the recurring, often serious problems they encounter in working with one another.

We bring to the task of describing the characteristics of effective partnering an understanding both of the partnering process and of the organizations and people that participate in it. Along with that understanding, we bring a strong interest in making partnering as effective as possible. We know partnering can work well and produce results beyond many participants' expectations. We want to see partnering evolve in a way that realizes its fullest potential.

Our specific thoughts and recommendations on improving partnering effectiveness run throughout this book. We begin here with several more general descriptions of what it takes to make partnering effective:

- *Effective partnering depends on commitment, thought, planning, and follow through.* It is difficult enough to hold an effective workshop, more difficult still to implement the workshop's results on the job. Effective partnering dedicates more effort to follow-up than to preconstruction workshops.

- *Effective partnering need not cost much money or time.* Effective partnering is a wise investment of effort in improving the extremely complex, costly, and time-consuming task of project communications. People with extensive experience in their own profession or trade do not always make realistic estimates of the need for or value of time devoted to improve communications.

- *Effective partnering is not a workshop.* Most partnering efforts use workshops, but effective partnering extends far beyond the workshop to influence communications on the job site, on the phone, and in organizations.

- *Effective partnering is not a "program."* Many partnering workshops cover similar topics, but effective partnering is never "canned." Effective partnering covers general concerns but also addresses specific project issues in depth.

- *Effective partnering is not alternative conflict resolution (ADR).* Partnering may use some ADR techniques, but partnering begins long before conflicts have even begun to form and reaches deep into the design process itself.

- *Effective partnering is not "soft."* Partnering does not encourage compromise or giving things up but rather focuses on communicating in a collaborative way that benefits all parties and the project as well. People argue when there is partnering, but they argue more skillfully and with more creative, productive results.

We offer this manual on partnering to readers with the same sentiment that might accompany a computer manual. Like a computer, partnering is a potentially valuable tool and resource. We want the manual to help readers understand how partnering works and to use that knowledge to derive the maximum possible results and value. The following chart reflects our philosophy as set forth in this book.

Our Opinions That Frame This Book

Partnering

- Partnering can be very effective
- Partnering can influence design and construction core practices
- All partnering efforts are not alike
- Some partnering efforts are more effective than others

Effective partnering

- Depends on commitment, thought and *follow-through*
- Need not cost much money or time, it is an *investment*
- Is not just a workshop, it is what happens *afterwards*
- Is not a "program," it is *customized* to every job
- Is not "alternative dispute resolution" (ADR), it is *more comprehensive and proactive*
- Is not "soft," it emphasizes *collaborative* problem-solving

---

PART 1

Partnering Core

What is partnering? What does it do? How does it work? What can I do to design a partnering effort for maximum returns? How can I get the most out of partnering as a participant?

This section provides a basic understanding of partnering. Specific chapters address the content, variations, specifics, benefits, deeper impacts and problems of partnering.

1

Partnering Content, Applications, and Results

Core Partnering Tasks

Partnering is a formal program to improve communications among the people and organizations working on a design and construction project. Key project team members convene for a preconstruction workshop and for regularly scheduled follow-up workshops over the course of the project. They often work with a facilitator to improve the quality and productivity of their discussions.

At the preconstruction workshop participants write, agree on, and plan to implement three documents:

1. *Goals statement*, which describes their hopes for the project.
2. *Communications procedures*, which specifies in detail how, when, and what people will communicate with one another on the project.
3. *Conflict resolution process*, which anticipates that people on the project will differ and maps out steps to help them resolve their conflicts to mutual satisfaction.

In follow-up workshops, participants:

- Address current concerns that have arisen on the project, using the workshop as a forum for resolving the inevitable conflicts and miscommunications that occur on a project.

- Review their performance according to the goals they set previously and develop action plans as needed.

- Tune and refine the communications procedures and conflict resolution processes they initially formulated in order to keep them effective.

Improving Partnering Effectiveness: Two Additional Tasks

It is one thing to meet in a workshop and write inspirational documents, but it is something else to use those documents to guide everyday project communications. Many partnering programs stop at the conclusion of the partnering documents and procedures. In order to improve partnering impact and effectiveness, it is important to go beyond these initial steps to include two additional tasks: communications skills training and direct involvement with the infrastructure of communications on the job site.

Without some work with improving communications skills, project team members will likely find it difficult to implement the partnering procedures they have devised. In addition, the procedures often imply a style of everyday communications that project team members may aspire to but in which they lack experience and skill.

Thus, to ensure that project team members have the communications skills necessary to implement partnering and to set a tone that supports the quality of the work, facilitators often provide project team members with communications skills training, focusing especially on:

- Conflict resolution skills

- Listening skills

- Valuing differences and diversity

- Creative problem solving

Design and construction projects must effectively manage a complex flow of information and communications among project team members. To improve its effectiveness it is very useful for partnering to strengthen the infrastructure of project communications.

One practical but crucial way partnering can strengthen the infrastructure of project communications is to improve the productivity of the numerous meetings that occur on the project. This involves providing participants with new skills and direct assistance in planning agendas and discussions.

A more technologically advanced and equally important way partnering can strengthen project communications infrastructure is through the use of groupware computer software. On the cutting edge of communications technology, design and construction projects are beginning to use such software. As computers and computer networks become more widely available and less expensive, partnering can include the development and implementation of "groupware." Groupware enhances partnering by providing an electronic means for carrying out many of the communications procedures that project team members attempt to address manually.

Overall, it is possible to categorize partnering activities into three categories: strategies, skills, and infrastructure. See Table 1.1.

Why Partnering Is Necessary

To many, the need for and benefits of partnering are obvious. Others wonder why, after so many years of getting along without it, partnering is necessary now. Partnering is necessary because it addresses a number of serious, complex problems that have become more pressing in recent years.

- *Litigation.* As in many businesses, costly litigation plagues many design and construction projects. Partnering can provide the structure, the skills, and a forum to resolve conflicts easily, before they get serious enough to become lawsuits.

- *Project costs.* Ineffective project communications directly impact project costs and the profitability of the firms working on the project. Few architecture firms make a profit on the construction phases of a project, usually because of ineffective communications. Engineering and construction firms, subcontractors, and the client also frequently lose money because of ineffective project communications, resulting in rework, mis-

Table 1.1   Core Partnering Tasks

Strategies	Skills	Infrastructure
Goals statement	*Training to improve:*	*Methods for organizing and documenting communications* (often devised in the communications procedures) including:
▪ What are project team members hopes for the project?	▪ Trust and mutual understanding	
	▪ Valuing differences	
▪ How do they want to communicate with one another?	▪ One-on-one communications	
	To set a tone in everyday communications	▪ Request for information (RFI) management
Communications procedures		
What specific procedures are necessary in order to attain the goals?	To enable people to resolve issues on their own	▪ RFI logs
		▪ Meeting minutes
		▪ Meeting agendas
Conflict resolution process	▪ Conflict resolution skills to help people resolve their own conflicts	▪ Daily on-site meetings
How, specifically, will people resolve inevitable conflicts?		▪ Senior organizational meetings
	▪ Group communications skills	▪ Documentation methods
	For running meetings	▪ Everyday project communications methods and tools
	For creative problem solving	
		Groupware computer software

takes, low quality, and lack of coordination. Partnering provides tools to improve the communications aspects of a job and reduce the resulting costs.

▪ *Fragmentation of the design and construction industry.* The design and construction industry has traditionally been rigidly divided into different trades and professions. In recent years these divisions have become even stronger and more finely delineated as some firms find economic rewards in specialization. New technologies have further fueled this trend.

Dividing tasks into small pieces results in some efficiencies but also can pit the pieces against each other. The design and construction industry has worked much more at dividing itself than in consolidating its fragmented pieces. Every project faces problems, the solutions of which could best come from concerted, coordinated efforts of the divided trades and professions.

Partnering can provide a forum for that coordination, bringing the different players together to address project-wide problems.

- *Complex, changing client organizations.* In their efforts to trim costs and improve productivity, many client organizations are undergoing far-reaching cutbacks and changes. These changes, particularly striking in large corporations and government agencies, make for a problematic client in design and construction. People and responsibilities change, continuity breaks, momentum stops.

  Partnering does not eliminate client change but it does help the project team manage client involvement by creating procedures, policies, and practices that can weather client organizational turmoil.

- *Design, construction, engineering, and real estate organizations face new challenges.* As clients become more productive, design, construction, and real estate organizations must keep up if not stay ahead. This is a challenge for firms and organizations that historically have lagged in using effective cost management and productivity tools.

- Partnering provides participating firms with in-depth exposure to new management ideas and methods. Partnering also can bring to light internal issues in participating firms, thus providing some insight into key internal issues to address.

The benefits of partnering are summarized in Table 1.2.

What Partnering Does: Clarifying and Managing Communications

Two subsequent chapters of this book explore in detail the benefits and impacts of partnering for the project, the project team members, and the client. Here it is important to describe more briefly how partnering works and what partnering does.

Table 1.2   Why Partnering Is Necessary

Issue	Partnering impacts
Litigation	Partnering provides strategies, skills, and a forum to address and manage conflicts before they become serious enough for litigation
Costs of poor communications: ■ Rework ■ Coordination ■ Poor quality	Partnering provides tools to improve communications effectiveness
Task fragmentation in the design and construction industry; some project problems cross trade and professional lines	Partnering provides a forum, a structure, and skills to address project-wide problems and issues
Changing, complex client organizations, especially large corporations and government agencies	In its methods, procedures, policies, and documents, partnering provides continuity to weather client organizational change
Design, engineering, construction, and real estate organizations face new challenges	Partnering can shed light on important internal issues and provide in-depth exposure to new ideas and practices

Partnering clarifies and manages project communications. Partnering manages project communications in much the same way that people already manage project scheduling and technical, craft, and professional tasks:

■ *Identify and clarify.* Project management involves getting project team members to identify and clarify what they are going to do and when they are going to do it so that others can plan and schedule their own work. Partnering involves getting people to identify and clarify their information needs and requirements.

■ *Commit.* Project management involves getting people to commit to task completion dates and levels of quality and effort. Partnering asks project team members to commit to carrying out tasks of communication.

- *Coordinate.* Project management involves coordinating the task efforts of different members of the project team. Partnering involves coordinating their communications.

- *Control.* Project management is an effort to control the overall progress and quality of the project. Partnering attempts to exert some control over project communications.

Partnering also differs from project management in some very important ways, stemming from the fact that partnering involves managing communications:

- *Intangible.* While project management results in products that are tangible, communications is an intangible. One of the things partnering does is to make communications more tangible in the form of procedures, processes, and documents.

- *Subjective.* In project management it is clear whether a task has been completed on time or not. The quality of communications, however, is much more subject to individual interpretations and differences. Partnering does not eliminate subjectivity but attempts to get people to be clearer and more specific in their use of terms that can be interpreted in different ways.

- *Group control.* While one person or one firm often has the responsibility for coordinating tasks and scheduling in project management, the project team as a group functions as the decision maker in partnering. Project team members take responsibility and decide for themselves how they want communications to take shape on the project.

Table 1.3 summarizes how partnering works and what partnering does.

Is Partnering Really Necessary? When Is Partnering Useful?

The American Institute Of Architects, the Associated General Contractors, and the American Consulting Engineers Council all endorse partnering. Some federal, state, and city agencies include partnering as part of the standard requirements in the bid package on large projects. Several state legislatures are considering

Table 1.3 What Partnering Does: Managing Project
Communications

Similarities to project management— both partnering and project management:	Differences from project management
Identify and clarify information	Partnering works with intangibles and subjective issues
Elicit and clarify individual commitments	Partnering commitments are voluntary
Coordinate individual action	Partnering involves the group in coordinating itself
Exert more overall control on the project	Partnering controls come from the project team

requiring partnering as a part of any new projects. Clearly, then, many people think partnering is worthwhile.

Instead of attempting to answer the more ideological question of whether partnering is necessary, we address the more specific issue of when partnering is beneficial. Conditions that make partnering especially beneficial follow from the preceding section.

"When is partnering useful?" is really a cost–benefit question. In other words, when do the benefits of partnering justify the costs? It is also more productive to ask "How much and what kinds of partnering will be useful?" rather than the more general question "When is partnering useful?" The more specific question can help design partnering to meet the needs of a specific project.

Subsequent sections of the next chapter, "What Partnering Does: Clarifying and Managing Communications" and "Types of Partnering," provide the foundations for answering the question "When is partnering useful?" Knowing what partnering does and learning about the various kinds of partnering efforts make it more possible to determine when it will be worthwhile to initiate a partnering program.

Some people begin with the questions "Will partnering solve all of the project's problems?" or "Can partnering change the personalities of the people on the project team?" or "Will partnering eliminate problems?"

While the answer to these more global questions is a resounding "no," the answer to the question "When is partnering useful?"

is "most of the time." After all, determining whether partnering is useful really reduces to a simple time-and-effort calculation: Does the time and effort spent on partnering pay for itself in improvements in the project?

When project budgets get up into the hundreds of thousands of dollars, and the potential costs of conflict, delays, and miscommunications keep pace, the cost of a few workshops is minuscule in comparison.

A final factor in determining the cost–benefit ratio of partnering involves understanding what partnering actually costs. Two factors contribute to the costs of partnering:

- *Direct costs of workshops.* These expenditures include a conference room, the cost of materials (from $0 to $30 per person), and a facilitator. Professional facilitators typically charge between $1,000 and $3,000 per day depending on their level of experience and skill. (Many facilitators also conduct and charge for between a half-day and a day of preparation and/or interviewing.)
- *"Contributed costs" of participants' billable time.* Usually these costs add up to a sum considerably larger than the direct costs of the workshop.

Overall, then, the total direct costs of typical workshop can be in the range of $2,000 to $4,000, a sum that is minuscule when compared to project budgets or to the potential costs of conflict or miscommunications.

Beyond these general considerations, we have identified nine different situations in which we have found partnering to be worthwhile:

- *Large client organization.* A large client organization, e.g., a large corporation or a government agency, often means different players will be involved in design and construction. Partnering can clarify who has responsibility for what and can provide procedures for occasions when people contradict one another.
- *Multiple clients.* Projects have multiple clients when one organization pays the bills but another organization occupies the building, or when several organizations pay and occupy. Government work often involves multiple clients, often with unclear authority and responsibilities. Whenever a project has

multiple clients, partnering provides a forum in which communications tangles and overlaps can be resolved.

- *Client in transition.* Client organizations that are getting restructured, reengineered, downsized, or otherwise changed pose several communications challenges. Partnering helps clarify some of the client's internal ambiguity and provides continuity of procedures when or if jobs and responsibilities are changed.

- *Politicized client.* If people in the client organization are jockeying for position, status, and power, their battles will be reflected in project communications. Partnering helps clarify the issues and declare a truce around the project.

- *Project team member organizations that are large, in transition, or politicized.* Whenever any other members of the project team (architect, engineer, contractor, subcontractors, etc.) are organizations that are large, politicized, or in transition, partnering can be helpful.

- *Government involvement.* Government design and construction personnel have wrongly earned a bad reputation. The real villain in government involvement in design and construction is usually not the people but the legacy of unwieldy procedures and rules they must live with. If government agencies are involved in a project in any way (as potential regulators, funders, tenants, or clients) partnering can help clarify the ambiguity and soften the inflexibility the agencies often (are forced to) represent.

- *"Forced marriages."* Many projects bring together firms that are unfamiliar with working with one another, and many of these forced marriages work out well. Some, however, bring together firms whose work methods, people, and/or styles of doing business are incompatible. Partnering can help parties work out the specifics of a working relationship.

- *Complex design, new technologies, and/or a difficult project.* Sometimes even a small project becomes a candidate for partnering if the work involved is difficult or complex. These projects demand more communications and more problem solving in order to translate the design into reality. Partnering can help provide the strategies and skills to strengthen the communications necessary for effective problem solving and information sharing among firms involved with the project.

- *Highly visible projects.* Some workers like an audience, others find that being watched increases the pressure of work that is already difficult and causes them to make mistakes they would not otherwise make. When construction projects are watched, either by the public or by a corporate client, errors may occur. If the public or the client organization has a high level of interest in a project, partnering can help in two ways:

It provides a means to make internal communications and problem solving on the project as effective as possible.

It manages the flow of information outside project team members. Partnering can help the team sort out who communicates what and when to the outside constituent. If the project team thinks it would be useful, partnering also provides an organized, controlled forum to bring the outside constituents into the project so they can see for themselves what is going on and thus participate in a manageable way.

See Table 1.4 for a synopsis of situations in which partnering is worthwhile.

Table 1.4   When Is Partnering Useful?

Wherever partnering provides a forum to surface and clarify ambiguity and confusion in client organizations:
- Large client organizations, e.g., corporations and government agencies
- Multiple clients
- Client organizations in transition that are going through downsizing, reorganizations, culture change, etc.
- Politicized client organizations

When any other member of the project team (architecture, engineering, contracting firms) fits any of the above descriptions

When government agencies are involved (as client, builder, or regulator)

When the contract creates forced (and unlikely) marriages of firms on the project team

When difficulties arise due to a complicated design, new technologies, or a complex project

When the project is highly visible in the community, professions, or trades

2
Common Themes and Variations in Partnering Practices

Summary

People in design, construction, engineering, and real estate use the term "partnering" loosely to describe a wide range of activities, including:

- Joint business ventures and real estate developments
- Informal attempts to cooperate and resolve conflicts among individuals
- Collaborations between public and private sector organizations
- Efforts to improve customer and client satisfaction

The *partnering* this book explores differs from these activities in that it is

- More organized, often as a program of specific meetings
- Proactive, often scheduled before construction begins
- More structured and planned

- A distinct set of strategies, skills, and tasks
- Linked with specific project outcomes and results

This chapter explores the core characteristics and misconceptions of partnering and examines different types of partnering.

Core Partnering Characteristics

Partnering varies widely from one project to another, yet there is also extensive consistency among most partnering efforts. These core characteristics of partnering are as follows:

- *Communications focus.* Partnering focuses on communications among project team members, not on their technical skills or professional expertise.

- *Inclusive.* Partnering brings together all the key players on a project in an effort to bridge many of the gaps set up by the way the design and construction industry divides the work.

- *Public.* Partnering brings people into the same room at the same time, rather than having communications follow a sequential chain among people in different locations.

- *Proactive.* Partnering attempts to take an active stance to managing many of the predictable (some would say inevitable) miscommunications and conflicts that arise in design and construction.

- *Cooperative.* In all partnering efforts there is an attempt to cooperate, collaborate, and get along well.

With this consistency, the ways in which partnering efforts may vary include

- *Built in or add-on features.* Whether partnering is included as an aspect of project communications from the outset, planned and listed in bid letters, or if it is added after the project has started.

- *Emphasis.* Which of the partnering tasks listed above are emphasized, neglected, or even ignored.

- *Execution.* Specifically, how the partnering tasks are carried out. For example, it is possible to rush a project team through writing a *goals statement*. Or it is possible to write a goals statement slowly and thoughtfully and to devote in-depth effort to planning how to implement the statement so that it impacts everyday work on the project.

- *Time.* How much time is devoted to the partnering effort. For some projects, partnering involves workshops that last several days. For others, partnering may not involve a workshop at all but simply a structured meeting of several hours duration.

- *Structure.* How the partnering effort is structured and organized. Sometimes people conduct partnering activities in intense one- or two-day workshops separate from everyday design and construction work. At other times people divide partnering activities into smaller tasks and link them with ongoing project work, conducting partnering tasks on the job site. The core partnering characteristics are summarized in Table 2.1.

Misconceptions: What Partnering Is Not

Whenever different people participate in the same activity, inevitably they will call it by different names and labels. At some point it is also inevitable that some of the evolving activities and assumptions people connect with the original label deviate significantly from the original idea.

The mistaken notions people have about partnering trace back to their experience with other, quite different innovations in the design and construction process, to other organization improvement efforts, and to their own imaginations. (See Table 2.2.) Some

Table 2.1 Partnering Core Characteristics

Core characteristics	Areas of variation
Communications focus	Built in and/or add-on features
Inclusive	Execution
Public	Emphasis
Proactive	Time
Cooperative	Structure

Table 2.2   Clarifying Myths and Misconceptions: What
Partnering Is and Is Not

Partnering truths	Misconceptions
Parterning *clarifies* and *manages* the communications of a project.	"We are partnering. We just had lunch together. Isn't that enough?"
Partnering *uncovers problems* and provides structure and skills to address them.	"Partnering is 'design by committee.'"
Partnering asks participants to *commit* to individual tasks to resolve issues.	"Partnering means I have to go along with the majority."
Partnering attempts to *coordinate* aspects of the highly fragmented nature of design and construction.	"Partnering is really just the way we always used to do business, in the good old days."
Partnering helps project team members work together to *control* more of the overall project and get to know enough about each other to increase trust.	"The formality of partnering will just get in the way of building trusting relationships."
Partnering includes alternative dispute resolution (ADR), but only as one of a number of strategies to improve communications.	"Partnering is just another label for ADR methods."

of the common misconceptions of partnering we encounter
include the following:

- *"Of course we're partnering, we just had lunch together. Isn't that enough?"* Informal efforts to get along well and cooperate, handshake arrangements, and a sense of humor all help build strong working relationships but they do not constitute partnering any more than looking at a calendar constitutes project scheduling or checking the firm's bank balance constitutes project budgeting. Partnering is a balanced set of more formal strategies and skills.

- *"Is partnering the same as 'design by committee?'"* The wave of interest in the 1960s in developing a more participative design process left a number of architects (and engineers and contractors as well) with a lingering distrust of anything that tampers with the design process. Partnering does not give the design process over to a committee. Partnering does recognize that many people need to be heard in the design process, so it tries to allow them to be heard in an organized way.

- *"Does partnering mean I have to go along with the rest of the project team, even if I think they are wrong?"* Conflict and disagreement are potentially powerful resources in solving problems. If it is handled well, disagreement keeps discussion open, stimulating, and creative. Partnering attempts not to eliminate conflict but to enable project team members to manage conflict effectively.

- *"Isn't partnering really the way we used to do business years ago, when people trusted each other, before the lawyers came onto the scene?"* Partnering does attempt to regain the kind of trust and strong working relationships people in the design and construction business remember from decades ago (not everyone, however, has these fond memories, we have discovered). At the same time, partnering also recognizes that projects are different now, being more complex and requiring different tools. Partnering aims to provide some of those tools.

- *"Partnering is just another name for those alternative dispute resolution (ADR) techniques that are popular now."* This is partly true, in the sense that partnering usually includes an approach to conflict resolution similar to those used in ADR methods. However, partnering includes much more work at the beginning of a project and throughout to help anticipate and avert conflicts before they need to be resolved.

- *"Isn't real 'partnering' just an understanding, a feeling of trust? Don't all the activities and formalities of partnering programs get in the way of people building trust in one another?"* If the design and construction business involved only individuals working for one another, partnering might not be useful. Increasingly, however, design and construction work involves organizations working with other organizations for a group of other organizations. In any kind of work with organizations, trust among individuals is not enough to make things work because the individuals are anchored in a complex organizational setting.

Partnering provides tools to manage not only the other organizations on the project team, but one's own organization as well. The formality of partnering helps provide the space and structure for individuals to build trusting relationships.

Typical Partnering Applications

With all the specific ways in which partnering can vary, overall, partnering applications can differ significantly from one another. Although all kinds of variations are possible, partnering typically takes shape in one of four different approaches:

- *Comprehensive partnering,* starting with the bidding process and following through to completed project. A variation of comprehensive partnering is *condensed partnering,* which uses all the tools of partnering but scales them down to use on smaller projects.
- *Problem-solving partnering,* which uses partnering methods to address project problems that arose after construction started.
- *Relationship building,* in which there is no specific project or problem, and service provider and client use partnering methods to strengthen their ongoing working relationships.
- *Piecemeal partnering,* in which people on a project use partnering methods outside the context of larger goals, priorities, and expectations.

See Table 2.3 for descriptions of the types of partnering.

Comprehensive Partnering

Because this kind of partnering begins prior to construction, with the building of partnering program provisions and design into the initial project *requests for proposals,* this kind of partnering usually starts with the client. With partnering provisions thought through for the whole project, people bidding on the project know, from the outset, that partnering will occur and how it will fit into the project. (See Fig. 2.1 for an example.)

Table 2.3 Types of Partnering and Their Outcomes

Type of partnering	Who usually initiates	When initiated	Focus	Outcomes
Comprehensive, including condensed partnering	Client, owner, government agency, corporate real estate organization.	Preconstruction, often in bid letters and requests for proposals.	Goals statement, communications and conflict procedures, training.	Overall improvement in communications, quality, productivity.
Problem solving	Project team member impacted by, responsible for, or involved in a specific conflict.	When a specific problem becomes clear and significant.	Solve problem at hand to a level of mutual satisfaction, stay out of litigation.	Specific problem resolved.
Relationship building	Service provider involved in ongoing client relationship or client interested in improving the quality of service.	When service provider or client senses relationship is deteriorating. Usually triggered by recurring mistakes.	Communications systems and procedures. Realign goals. Renew contact between people.	Improved procedures and systems. Strengthened relationships. Improved problem solving.
Piecemeal partnering	Anyone on the project team who has the authority to initiate a "partnering" task.	Whenever people are moved. Often in response to a problem.	Driven by the interests of the person initiating the activity.	Possible results from specific activities; possible frustration from limited successes.

Jim Donovan cleared his desk to make the space to organize all the brochures, handbooks, and notes he had accumulated on partnering. He wanted to make a clean start, to have everything he would need in front of him if he was going to conduct a useful pilot of partnering for the State Building Commission. Jim had been to enough seminars on partnering now and participated in one partnering workshop for another agency. He felt comfortable with his own understanding of how partnering could work for the Commission.

Jim planned it the right way, from the ground up. He would include information about partnering right in the request for proposal (RFP) so that contractors would know from the very beginning what to expect. Actually, Jim knew that most of the contractors he worked with had already been involved in partnering in some way, so he did not expect them to be at all concerned about the partnering requirement. If anything, they would probably be relieved. Pleased with the returns they had experienced from partnering on other projects, several of the contractors had been nagging Jim for some months to get the Commission to use partnering as well.

Even with this kind of support, there was still a lot to figure out. What should the RFP say? How much detail did Jim need to provide on partnering?

Jim knew there should be a preconstruction workshop, but he did not know how long it should be: one day, two days? What should be on the agenda? Who should be invited?

Jim also knew there would have to be follow-up meetings, but again, he was unsure of the details. How often should the follow-up sessions occur? Was it really necessary to have a whole workshop to address issues? Could they incorporate follow-up discussions into regular project meetings?

There was a lot to figure out, Jim repeated to himself, but he knew it would be worth working on. Once the Commission got some experience with partnering, it would really help him stay in touch with the projects he worked on. It would help the agency get better buildings. It would help the agency's service providers communicate more efficiently with the agency.

Figure 2.1. Comprehensive partnering: A profile.

It would take a good plan to get the partnering pilot off the ground, Jim knew. This was the state government he was dealing with; there could be no surprises. Jim would need a good plan, a thorough plan that laid out all the options and all the possible courses of action right from the beginning of the building to the opening ceremonies.

Figure 2.1. (*Continued*)

Usually comprehensive partnering includes at least a preconstruction workshop and a follow-up effort of some kind and also includes all the partnering tasks listed above: goals statement, communications procedures, and conflict resolution. Comprehensive partnering is the most planned and proactive of all types of partnering.

Condensed parterning is an important variation of comprehensive partnering that is popular and effective with many corporate real estate departments, government agencies, and other organizations that do large numbers of small projects. They use all the elements of comprehensive partnering but reduce the time and effort involved in order to match the reduced scope of the project.

In these situations, there may be no workshop. The contracting officer in the real estate or design and construction department may reduce partnering to a meeting of a few hours with the client and one or two other key players on the project. However, the contracting officer still covers all three of the major components of partnering (goals statement, communications procedures, and conflict resolution). In other words, condensed partnering is not to be confused with "piecemeal partnering," described later in this section.

Corporate real estate and design and construction departments that use this method report that it improves their clients' level of satisfaction with the services they provide. In addition, they say that this approach helps them to improve a chronic problem of working with large organizations: clients who have limited understanding of what design and construction firms do and clients who take limited responsibility for their own involvement and the costs of last-minute changes. In short, these agencies say that partnering helps them manage their clients.

Problem-Solving Partnering

On a scale of proactive to reactive, problem-solving partnering is at the reactive end. Any of several project team members can suggest and initiate partnering when problems affect a project. Usually the project team member who stands to lose the most is the person who makes the suggestion and drives the effort.

Problem-solving partnering usually focuses on resolving several (there are usually more than one) problems that have come up on a project, so it places a great deal of effort on resolving specific conflicts. Sometimes, once project team members are together, they may also work on a goals statement, communications procedures, and a conflict resolution process as a way to prevent similar problems from recurring. (See Fig. 2.2.)

Partnering as Long-Term Relationship Building

Sometimes clients and firms that have a long-term working relationship with expectations of continuing work together will participate in partnering in order to strengthen that relationship. Although it is possible for client organizations to initiate this kind of partnering, we have more often seen firms initiate it as a way of building client satisfaction.

Drawing from projects currently in the works, recently completed, and projected in the future, this type of partnering tends to focus on producing procedures and systems for managing communications for maximum mutual benefit. (See Fig. 2.3.)

Piecemeal Partnering

With the proliferation of partnering programs and the diversity of what those programs actually involve in terms of activities, tasks, and skills, it is inevitable that people working on projects will put their own efforts together, piecemeal, under the "partnering" banner.

A project manager may have participated in a comprehensive partnering program but has the power on his project to experiment only with a goals statement. Or a contractor might label as "partnering" a simple meeting. For example, the contractor might

George Clark reached for the phone hesitantly. On the one hand, he did not relish making this call, admitting his firm's mistakes and delays to the contractor. On the other hand, he had no one else to go to. Besides, it was at least partly the contractor's fault that the building was falling behind schedule. True, George's design firm was slow in getting requests for information (RFIs) back to the contractor, but the contractor's people didn't fill the reports out correctly in the first place most of the time.

Frank Higgins enjoyed hearing George eat a little crow about the delays, and he faked a pretty good argument about being in the right when it came to filling out RFIs. After all, Frank's firm had built its excellent reputation on on-time scheduling and completion of many challenging projects. He knew he had the edge on George, and he enjoyed it.

Frank also knew all too well, though, that his own workers were much less than perfect in the field when it came to things like RFIs. Frank had to fight with the workers every week to get their time cards in. If they had trouble filling out their own hours to get paid, Frank knew they would have trouble communicating about abstract problem solving to an architect.

Frank listened to George go on for a few minutes and then suggested that they try to settle what they could in a partnering workshop. At first George held back. He was concerned that there were already too many meetings going on, and he had hoped that Frank and he could settle the whole thing maybe just with a lunch meeting.

Frank held his ground; he wanted a bigger meeting. There were too many people involved now to make it possible just for George and him to handle it by themselves. If some of the men in the field could sit in on the meeting, it would give them a much better sense of how much trouble they were causing. Similarly, if some of the other staff members in George's office could sit in, it would help them see how much difficulty they caused when they "lost" one of the RFIs that came into the office.

Frank had attended some partnering workshops for other jobs his company worked on, so he made some suggestions to George about what should go on the agenda. First, they should

Figure 2.2. Problem-solving partnering: A profile.

try and get to the heart of this RFI tangle. They should sort out which ones were on time and which were delayed and put together some kind of procedure to ensure the delays would stop.

George felt more positive about partnering. Once he came to terms with the necessity of the meeting, he found that he had several other suggestions for topics that could go on the agenda:

- The bankruptcy of the electrical contractors impacted everyone, he knew. Maybe it would be useful to discuss how the other subcontractors would have to alter their original plans in order to accommodate the delay.

- Town building inspectors seemed to be pitting different contractors against each other. Maybe there could be some way to coordinate their visits and the procedures they had to follow.

- New technology had become available for some of the electrical work he had designed. George was curious if any of the other members of the project team had had any experience with the new technology, and wondered if they thought it could be used on the project.

Frank agreed with George that these would be useful items to place on the agenda and added a few items of his own. Then they reviewed the project team list and divided it between the two of them; George would call half of them, and Frank would call the other half. Together they would see if they could get the problem-solving partnering off the ground.

Figure 2.2. (*Continued*)

bring all the project team members into the same room to try and resolve some communications problems without using any formal conflict resolution methods.

We have mixed feelings about these piecemeal "partnering" efforts. On the one hand, they usually reflect good intentions and can do little harm. On the other hand, they don't *always* reflect good intentions. (For example, the contractor in the situation

"You want to have a partnering session at a time like this?" Carol Scanlon, Megacorp's corporate real estate manager, was surprised at Jack Palmer's suggestion. The company was changing, cutting back. No one's job was secure. Everyone in the department had been moved around in the past year, and they would probably be moved again in the coming year. Why do we need partnering when we're not sure what we're going to be building?

Jack Palmer wanted to have a partnering workshop precisely because of all the uncertainty. He thought that if he could get all the major players into the same room at the same time, he might be able to get some answers to the questions that were delaying his firm's projects.

Having Megacorp for a client had been a mixed blessing for Jack's firm. They gave the firm plenty of work, treated Jack and his staff well and helped the firm learn a lot about perfecting the kinds of buildings Megacorp often built. Other firms in town envied Jack's firm's apparent financial stability.

On the other hand, Megacorp tended to swallow up Jack's firm's best efforts. Sometimes it felt more like they worked for Megacorp than for the firm. Also, whenever Megacorp went through any kind of instability, Jack's firm felt the aftershocks.

Like now. Megacorp was doing one of its cutback routines. Jack had seen it many times in the past, as the company tried to trim costs and gain productivity. Management by objectives, layoffs, cutbacks, downsizing. This time they were calling it "rightsizing."

Usually Jack didn't let Megacorp's wheeling and dealing trouble him, but this time was different. It was bad enough they had terminated some of the best people Jack worked with; he would miss them. Still worse, it was taking them months, nearly a year to get some clarity about who would be responsible for what. No one was stepping forward to take the reins on the half-dozen projects Jack's firm was working on. Six projects had been stalled for over a year, and it was costing the firm a great deal. Megacorp could survive something like this, but Jack wasn't so sure his own firm could.

As Carol Scanlon reviewed Jack's proposed agenda for the workshop, she could see that it would probably be very useful to have a partnering workshop. Jack's chief concerns were with

Figure 2.3.   Relationship-building partnering: A profile.

communications procedures on the half-dozen jobs his firm was working on. If they could get Megacorp's current project staff together, they ought to be able to clear up a lot of the delays that were bothering Jack. If they made progress on only one of the projects, the meeting would be worthwhile. It would communicate to everyone her expectation that people not wait until the company stopped changing. The way things were going, that was not likely to happen for years.

Partnering would be a good thing, Carol thought, because the delays were bothering her, too. She did not want to go down in the corporate record books as the first woman real estate director and the first person in the position to add nothing to the company's prestigious real estate portfolio.

Carol revised Jack's agenda a bit, added a few items of her own, and began making calls to her staff.

Figure 2.3. (Continued)

above may be convening the meeting in order to advance his or her own point of view.)

The piecemeal efforts at "partnering" can do three kinds of damage. First, by providing less than the full range of available partnering tools and methods, they may limit what can be improved. For example, project team members who write a goals statement without also working on communications procedures and conflict resolution are likely to have limited success with the impacts of the statement.

Second, participants' perceptions of what can be improved are also likely to be limited. People who have participated in a piecemeal "partnering" effort may fail in their attempts to improve communications because they have not been given the opportunity to address the full range of issues that will provide results.

Piecemeal "partnering" efforts also damage the viability of more traditional partnering by creating a negative image. People who participate in piecemeal "partnering" may conclude that partnering does not work when they haven't really tried it.

3
Specifics

Summary

This chapter responds to the common concerns people express about partnering and the most frequent specific questions they ask. The two general areas in which people express concerns are legal and professional:

- What does my lawyer say about partnering?
- What will my peers say about partnering?

The specific questions people ask include

- Who should be involved in partnering workshops and activities?
- When should workshops and activities be scheduled?
- Where should workshops and activities be held?

We conclude the chapter with a list of specific steps to take in planning and implementing partnering to improve partnering effectiveness.

Legal Concerns: What Will My Lawyer Say About Partnering?

Attorneys and partnering do not always mix well, which is understandable. Some attorneys advise clients not to participate in partnering. They worry that, in the openness of partnering discussions their clients may divulge information that could be used against them later if aspects of the project came to trial. They worry that by divulging such information, their clients may become legally vulnerable.

Some attorneys also worry that, in the course of partnering discussions, their clients may make commitments with which they cannot legally follow through and that are to their legal disadvantage. Attorneys fear that their clients may make commitments without fully understanding legal documents: the construction loan agreement and the bonding indemnity agreements, e.g.

These concerns of attorneys reflect legitimate worries. However, the lack of understanding many attorneys have for any kind of communications outside traditional legal disputes has detracted from their credibility in discussing partnering. With issues such as partnering that are not clearly, directly connected to a legal matter, clients are sometimes unwilling to pay attention to their attorneys, much less use them effectively.

Knowing their attorneys' preference for and comfort with legal disputes, many design and construction practitioners participate in partnering without their attorneys' knowledge. Some practitioners have taken an even more active stance against getting their attorneys involved in partnering, actually denying their attorneys' requests to participate in an advisory role in partnering workshops.

A general contractor comments, "I know that if I bring my lawyer to a partnering workshop, even if he just sits in informally, the architect and client will bring their lawyers too. Next thing you know, the lawyers will all be talking to each other, running the show. We the practitioners will be out in the cold, out of control."

The truth of the matter with respect to legal aspects of partnering is only a frustratingly current and transient truth, not a permanent one. The issue is rapidly evolving on both the legal and the design and construction practitioners' sides, with positions and roles changing and developing.

For attorneys, the classic role of the law as "enforcement of contracts" is rapidly becoming extinct. As technology, project complexity, and court costs all escalate, it is becoming increasingly difficult, nearly impossible, to determine and assign blame in the classic legal sense.

As attorneys realize this, they also face the fact that their previous approach and skills do not necessarily prepare them well for assuming a different role. Many attorneys have been trained and rewarded for arguing and debating in very much of a win–lose framework. The same skills and instincts that make them effective advocates may block their ability to learn to facilitate.

On the design and construction practitioner's side, it is not necessarily advisable to enter into partnering discussions without a full understanding of the implications for project communications of the project's legal documents. This does not necessarily imply that attorneys should attend partnering workshops, but it may mean that the practitioner should review the documents with an attorney before beginning work with partnering.

There may be some legitimacy to the attorney's expressed concern that people participating in partnering workshops may say something they live to regret later in a court of law. However, this is not a concern for which we can find any precedent. In addition, it is difficult to imagine any aspect of partnering discussions on real projects that involves information or commitments that make a participant legally vulnerable.

Professional Concerns: What Will My Peers Say about Partnering?

"Off to a partnering workshop? Be honest and tell us if it fails," their peers taunt as design and construction professionals depart for partnering workshops.

People who work in the design and construction fields often work for smaller organizations that do not have the kinds of internal training and development resources of larger corporations. Thus they may not know that the kind of group process work that goes on in partnering is quite common and typical. Large corporations typically use well-staffed internal training departments to teach managers and employees the communications skills people use in partnering. Moreover, many large corporations use internal organization development departments to conduct team-building programs that closely resemble partnering.

Table 3.1   The Partnering–Total Quality Management
(TQM) Connection

Both TQM and partnering:
Work within a philosophy of continuous improvement
Use similar tools
Work with similar skills
Use small groups to solve operational problems
Aim to improve relationships with customers
Depend for their success not on the technique itself but on follow through

   Design and construction practitioners may be comforted to
know that partnering includes practices, methods, and tools that
the larger corporate world has been refining for some years, tools
that yield predictably positive results.

   Partnering links especially well with internal organization
development efforts in *total quality management* (TQM) (see Table
3.1). Specific connections between partnering and TQM include

- Both work within a continuous improvement approach.

- Both partnering and TQM use similar tools: goals statements,
  mission statements, etc.

- Both partnering and TQM draw on similar communications
  skills training methods.

- Both partnering and TQM use small groups to address and
  solve everyday work problems.

- TQM often works within a framework of improving customer
  satisfaction with customers; partnering offers a set of specific
  methods for strengthening customer relationships. Partnering
  does with customers what TQM attempts to achieve internally
  in the organization.

- The key to effectiveness for both TQM and partnering is in the
  extent to which they extend the reach of program workshops
  back into participants' everyday work lives. When people say,
  "TQM didn't work" they usually are not commenting on the
  effectiveness of TQM as a tool, they are commenting on the

ability of their organization to follow through and implement the insights TQM generates. Similarly with partnering, effectiveness depends on follow-through.

Partnering Specifics and Logistics

Much of the success and impacts of partnering depends on how the client and facilitator handle specific logistical details, the "five W's" of partnering: who, what, when, where, and why. These emerge as issues of scheduling, timing, attendance lists, location, and purposes.

Decisions about these "specifics" constitute a major opportunity for improving partnering effectiveness, to use the options and alternatives in each of these areas most thoughtfully to design and deliver a program with maximum impact and results. Few of the specifics have consistently "correct" decisions. In making decisions for any one of them it is essential to have some understanding of the alternatives available and the implications of the different possible choices.

Who Should Participate in Partnering?

This is often the first question people encounter when trying to plan a partnering workshop. It may seem like a simple question, but it can be difficult to answer (see Table 3.2). We work with project groups ranging in size from 10 to 30. Factors to consider when deciding who should attend include the following:

- *Group problem-solving effectiveness.* When groups of any kind get to be larger than about 25 participants they become too large and unwieldy to function effectively as problem-solving groups. Large groups can listen and vote, but they are too cumbersome to engage in creative thinking.

- *Implementers and roadblocks.* It is important to include in partnering any project stakeholder who later may have the organizational authority to reverse or undermine the products of a partnering workshop. When we recently explained this crite-

Table 3.2 Who Should Attend Partnering?
Factors to Consider

- *Optimum size of group for problem solving*
  Groups larger than 25 can vote on an issue, but are too cumbersome for effective problem solving

- *Actively involve key decision makers and potential roadblocks*
  Anyone who could later block the actions of a partnering workshop should be actively involved or, at a minimum, involved in preapproving the partnering effort

- *Involve a project cross section*
  People high enough in their firms (owners, principals) to sign off on partnering agreements
  People actively on the job site to bring partnering agreements back to the job's everyday environment and communications

- *Representative from each of the major subcontractors*

- *Representative from tenant or occupant*

- *Facility manager*

- *Any other active stakeholder*

---

rion to a client who was working with a federal agency, he responded, "Then we might have 200 people at the workshop! This is a very tangled bureaucracy!" The point here is to try and anticipate where roadblocks may occur and to invite the appropriate people directly.

- *Project cross section.* The workshop group should include people placed high enough in their organizations to be able to make commitments on which their organizations can follow up. On the other hand, the group should extend far enough down into the project to be able to carry workshop decisions back to the job site.

More specifically, typical workshop groups number about 20 to 25 people for a $20 million project. Participants include

- Several people from each major organization: client, architecture firm, and general contractor. Each organization sends a senior decision maker (owner, senior principal, etc.) as well as one or two people working directly on the job.

- A representative from each of the major subcontractors.
- A representative of the organization that will occupy the building: facility manager, tenant group, etc.

When Should Partnering Occur? Scheduling Initial Workshops

Ideally, partnering should begin long before construction starts in the bid letters that go out for the project. By starting early, the major stakeholders can shape project team members' expectations for how partnering will be used and for what the project team members' obligations will be. Also by starting early, people can get to work quickly on whatever special or difficult issues may be impacting the project: special scheduling considerations, the use of new technologies, special site conditions, etc.

When partnering is a comprehensive effort, beginning in the bid documents it is useful to schedule the initial workshop several weeks to a month into construction. It is sometimes possible to schedule the workshop before ground is broken, and in some ways that may seem preferable. Starting partnering before any people communicate with one another makes it possible for partnering to shape the total project, right from the beginning.

On the other hand, starting partnering before the project begins runs the risk of making partnering abstract and unrealistic. People in design and construction, who are most familiar in dealing with the real world, sometimes become uncomfortable and ineffective when asked to deal with hypothetical questions.

In addition, workshops that occur before construction begins run the risk of having their output invalidated by project conditions. Often, people on the project team discover "surprise" project conditions and problems within the first month of construction. Leaving those factors out of initial partnering discussions makes it difficult for the partnering output to seem relevant back on the job site.

If partnering starts several weeks to a month into construction, it may not be able to shape every single communication from the ground up. However, starting a month into the job is still early enough so that partnering can exert a great deal of influence on project communications.

When Should Partnering Occur? Scheduling Follow-Up Activities

While partnering publicity comes in large part from initial workshops, overall partnering effectiveness depends much more on follow-up activities and workshops. Thus it is essential that the follow-up process be managed and scheduled so that participants can indeed follow up on the initial session and so that these workshops can provide a forum for new issues.

The heading of this section notes "follow-up activities" rather than "follow-up workshops" because much of the follow-up work of partnering does not occur in workshops but on the job site, in peoples' offices, or at morning meetings. To help summarize follow-up activities and provide some action focus to partnering workshops, we often begin workshops by posting an *action steps* summary sheet on the wall. Then, as people mention action items, we try to encourage them to commit to specific times, dates, and places. The action steps page often makes for a useful summary of what commitments people made at the meeting. See Table 3.3 for a sample worksheet on follow-up activities.

When Should Partnering Occur? Scheduling Follow-Up Workshops

Beyond specific follow-up activities, follow-up workshops can play a special role in keeping partnering effective. More than specific activities, workshops can provide a more comprehensive forum to take stock of project and partnering progress, to identify emerging issues, and to mobilize efforts quickly.

The general, recurring problem we encounter with follow-up workshops is that people do not schedule them often or soon enough. Part of the problem in this stems from a successful initial workshop. Once people have written a *goals statement, communications procedures,* and a *conflict resolution process* and have participated in communications skills training, it may be difficult for them to anticipate that they will need a follow-up meeting.

In addition, follow-up workshops encounter the same resistance that people express about the initial workshops, only more

Table 3.3 Partnering Follow-Up Activities
Sample Worksheet

Who	What	When and where
Site superintendent	Try to involve all participants actively	Weekly project meetings
Subcontractors and their respective area consultants	Meet prior to construction to review drawings	Within one week, at client's offices
Project architect and contractor's site superintendent	Organized a walk through of site twice weekly	Starting Tuesday, meet at 9:00 A.M. in trailer, e.g.
Project architect, architect's draftsperson, and contractor's site superintendent	Provide complete responses to as much as possible on site and do draft drawings	Starting Tuesday with project walk through, e.g.
Senior management of architecture firm, contractor, and client	Biweekly dinner meeting to head off evolving problems and discuss emerging issues	Starting in one week, at downtown restaurant

so. Some people do not see the need to meet at all; they assume that "things will work out" even when they are involved in litigation on another project. They participate in an initial workshop begrudgingly and then assume that they have paid their dues.

Sometimes people who actively support partnering also underestimate the need to follow up, more out of habit. They are just not used to taking a proactive approach; they are also not used to the idea of taking time to maintain work relationships. Sometimes people fail to follow up quickly or thoroughly enough because they picture the follow-up meeting as a carbon copy of the initial workshop.

Follow-up workshops need not necessarily fit the format of a two-day or even a whole-day program. On smaller jobs in particular it can be extremely difficult for project team members to take that kind of extended time off the job. Two half-day meetings, scheduled every two weeks, may enable project team members to use partnering to be apprised of project developments in a timely manner while also impacting their project schedules less dramatically.

On the other hand, it is also important to provide in partnering workshops time for in-depth problem solving that is not available in the natural course of regular project meetings. Thus a partnering meeting of just a few hours may not be effective in providing a meaningful forum to address difficult communications issues.

No matter when follow-up work is scheduled, one important principle to follow is to allow enough time at the end of any workshop to schedule the next session thoughtfully. When all the key project players are together in a partnering workshop it is much easier to schedule follow-up activities. Once people have returned to their offices or job sites it can take hours to find a common time for the next session.

As with other issues, a number of factors influence the scheduling of follow-up workshops. If the initial workshop set a number of important procedures in place, if upcoming project communications demands appear complicated, and/or if several people on the project team appear to have difficulty in communicating with one another, it is probably important to follow up quickly.

Specific factors that influence the most effective form, design, and timing of follow-up activities include the following:

- *Frequency.* More frequent meetings make it possible for the follow-up workshops to address new issues easily. However, increased frequency also takes time from direct project work.

- *Length.* Short follow-up workshops (one day or half day) make it possible to meet more often and maintain an action focus. Longer meetings provide more of an opportunity to address difficult issues.

- *Location.* While initial workshops usually are held off site, follow-up meetings may be more effective close to the site, even in the construction trailer. On the other hand, meeting in the trailer may make the session vulnerable to short-term distractions.

- *Participants.* Working on specific follow-up issues may make it useful to involve people who missed the initial workshop. On the other hand, new people may not easily fit in follow-up workshops without some preparation from others.

- *Biases and preferences.* People who are not comfortable with meetings of any kind may fail to schedule a follow-up workshop soon enough.

Where Should Partnering Workshops and Activities Be Held?

Many people assume that partnering workshops should be scheduled off the project site because they worry, and rightfully so, that holding the meetings on the site would leave them vulnerable to interruptions. They also worry that holding them in any one participant's office makes some kind of a statement that the host disproportionately "owns" the results of the workshop.

Both concerns are justifiable. Many efforts to save some project dollars by scheduling a partnering workshop in someone's office have left the participants so vulnerable to interruptions and distractions that they were unable to participate fully.

On the other hand, there are three circumstances in which it may be very useful not to hold workshops in a hotel or conference center:

- *Client's offices.* Holding workshops in the client's offices makes the statement that project team members are accountable to the client and helps keep the client in the project information loop. From a logistical standpoint, holding workshops in the client's offices also helps project team members make personal contact with people in the client organization.

- *Rotating project team members' offices.* Rotating partnering meetings among project team member's offices rotates ownership of partnering work while enabling project team members to make connections within one another's organizations.

- *Construction trailer.* The inevitable disruptions that occur when workshops are held in the trailer are often outweighed by the ease with which participants can link the workshop to real project concerns and solutions.

Using, Finding, and Qualifying Facilitators

Most partnering participants quickly agree on the need for an external facilitator. People working in each of the building professions and trades are usually aware of the strengths of their own biases and the need for someone with external objectivity to balance those biases.

There is also seldom a problem with covering the facilitator's fee. Usually the fee is in the range of 1 to 3 thousand dollars per day plus program materials, so the fee amount is small or even negligible in the overall scope of project costs.

The two more difficult issues to resolve for facilitators are finding them and qualifying them. The professional associations provide some guidance in locating facilitators. The national offices of the Associated General Contractors and American Institute of Architects provide names of facilitators, as do some of the local chapter offices of each.

No professional association certifies, ranks, or recommends specific people, however, so it is left to the people initiating and sponsoring partnering to make choices as best they are able. Many projects find it useful to begin partnering by convening a small group of people (typically owner, architect, and general contractor) to interview prospective facilitators.

Such interviewing seems like a logical approach to deciding on a facilitator but interviews can be misleading. Specifically, people with a great deal of design and construction experience but with limited experience in facilitation may appear more experienced than they are. Large corporations providing partnering consultation may appear more established in an interview but may offer more "packaged" approaches.

Thus it is often best in choosing a facilitator to go slowly and to check references extensively. A skilled facilitator will understand that the interviewers are using the interviewing task to pilot their initial partnering work.

Specific Steps to Improve Partnering Effectiveness

In keeping with this book's theme of improving partnering effectiveness, we conclude this chapter with a list of specific tips in scheduling, logistics, and program design that help improve the impact of partnering efforts (see also Fig. 3.1).

1. Partnering planning
   a. *Start before you start.* Workshops are more valuable when the facilitator interviews, perhaps even surveys, participants ahead of time. This helps the facilitator plan the agenda

1. Planning
   a. Start before you start.
   b. Use a cost–benefit approach.
   c. Make sure the right people attend.
   d. Share direct partnering costs.

2. Program content
   a. Be comprehensive.
   b. Focus on the request for information (RFI) process.
   c. Discuss implementation plans in detail.
   d. Anticipate long-term issues.
   e. Celebrate successes.

3. Partnering documents
   a. Sign.
   b. File carefully.
   c. Send out as follow-up material.
   d. Keep bringing to all meetings.
   e. Refer to them often.

4. Provide training
   a. Value differences.
   b. Acquire one-on-one communications skills.

Figure 3.1. Specific steps to improve partnering effectiveness.

while also kindling participants' thinking about their own involvement.

b. Plan the scope of partnering work with a cost–benefit approach. In trying to determine the extent of partnering workshops and activities, keep in mind what partnering can return in light of what it costs.

c. Make sure the right people attend. Determining who the "right" people are can be tricky on many projects, but this is not an issue to rush or gloss over.

d. Share direct partnering costs. Sometimes a project team member will volunteer to bear the brunt of partnering costs because they think they stand to gain from improved communications. While such volunteering is noble, it can undermine partnering effectiveness by limiting the extent to which other project team members own the results. Direct partnering costs are seldom significant. In fact, they are often trivial when compared to overall project budgets and the expected value of partnering outcomes. If it is at all possible, it is very useful for project team members to share in the direct costs of the partnering effort.

2. Partnering content

   a. *No matter when partnering starts, try to be comprehensive.* Sometimes when partnering starts after the initial phases of a project, it is tempting to omit some of the tasks, such as the goals statement. This is understandable but harmful to partnering effectiveness because it makes the partnering effort more tactical than strategic. Time taken to address the larger issues usually pays back in quicker, more effective resolution of specific issues later on.

   b. *Explore, clarify, and manage the request for information (RFI) process.* Many project communications problems revolve around the RFI process. From the outset, devoting concerted effort to managing RFIs usually creates larger returns for project communications.

   c. *Take time to think through and discuss implementing workshop ideas in detail.* It is far easier for a workshop group to come up with an idea than it is to implement it. Workshop time is well spent when participants discuss the fine details of implementing ideas: who will do what when, where, and how.

   d. *Take time in every workshop and meeting to anticipate long-term issues as well as to address short-term concerns.* Partnering participants often know but "forget" long-term concerns while they focus, understandably, on short-term emergencies. Listing "long-term project concerns" as the standing agenda item for every Partnering meeting, often helps a group remember and manage key long-term project issues.

   e. *Celebrate successes.* Participants' focus on short-term emergencies may help them "forget" some of the important successes logged by their own efforts at previous partnering meetings. This, again, is understandable but may detract from partnering effectiveness by limiting participants' confidence to keep bringing problems and issues to the partnering process.

3. Partnering documents (goals statement, communications procedures and conflict resolution process)

      a. Sign the documents. Sometimes people feel awkward asking others to sign the documents. It may seem like a small point at the time, but the impact of the documents later on increases significantly if all project team members sign them.

      b. File the documents carefully. Occasionally, in the positive feelings after a partnering workshop, no one recalls the necessity of taking and filing the documents. It is useful to agree, well ahead of time, who will handle the documents.

      c. Send copies of the documents (and signatures) to participants. Partnering documents, with signatures, provide useful follow-up activity and reminders after workshops.

      d. Bring the documents to all workshops and refer to them often. Taping the original documents to the wall in follow-up workshops helps maintain partnering continuity.

4. Training

      a. Value differences. Provide enough training in valuing differences so project team members can learn what they need to know about one another's thinking and communications styles to communicate effectively.

      b. Acquire one-on-one skills. Provide enough training in collaborative communications skills so participants can implement partnering documents and communicate in their everyday interactions to reflect partnering intentions.

4
Benefits: What Partnering Accomplishes

Summary

A contractor in the Midwest tells us he is working on two jobs for the same government agency client, both jobs about the same size and complexity. One of the jobs is using partnering; the other is not. Our man says the project with partnering is going better because of the partnering process. There is a noticeable difference in day-to-day project communications, both within and between the trades and among all the members of the project team.

The contractor explains, "I don't know exactly why or how it happened, but it is clear to me that the people on the job with partnering treat each other better and communicate more effectively. They are producing a better quality building as a result."

This is a nice story, but like too many anecdotes about partnering, it is too vague to be useful. To provide more useful information on partnering we explore first, in this chapter, its immediate benefits, both anticipated and unanticipated. We also explore how partnering benefits specific members of the project team.

In the next chapter we examine more closely how partnering produces these benefits. We explore its impact on factors that shape and define design and construction projects:

- The highly fragmented nature of the design and construction industry
- Individual project participation
- Design and construction organizations
- The nature of communications

Immediate Project Benefits

Participants in a partnering program describe its outcomes in several different areas:

- Reduced costs, usually as a result of compacted or more closely monitored project schedules and time lines
- Reduced frequency, extent, and severity of litigation
- Increased building quality
- Improved safety on the project

It is possible to link each of the impacts to specific components of partnering programs.

Reduced Costs. The chief means for reducing project costs on many projects is to cut back on the schedule for completion. If the building can be completed more quickly, contractors and subcontractors can trim their labor costs and move their people off to other jobs.

In partnering workshops, schedules and time lines become the subject of group discussion. The open discussion can result in schedule improvements for two reasons: improved monitoring and creative problem solving. Partnering workshops improve the monitoring of schedules by placing them under the public scrutiny of the group at the workshop. Problems become more visible and more open to suggestions for improvements.

At the same time, partnering programs may tackle an existing project schedule and attempt, through creative problem-solving techniques, to cut it back. Partnering workshops bring together the whole group of project players necessary to take a look at the schedule. When they see that they can work together in the partnering workshop, they can be encouraged to tackle larger problems such as the schedule.

Reduced Extent and Impact of Litigation. Fear of litigation is the motivation that drives many people to experiment with partnering, and with good reason. Litigation in design and construction has become a costly threat not only to the profitability of firms but to the viability of the design and construction industry overall.

In its pamphlet on partnering, the Associated General Contractors (AGC) notes:

> ...we have seen a dramatic increase in litigation, which is expensive and counterproductive to everyone's efforts to produce quality projects on time and within budget. The AGC strongly believes that the time has come for all the parties in the construction process to step forward and work together to take control of this costly and intolerable situation.

Extensive litigation grew to be intolerable in part because of the direct costs involved with working with lawyers, courtrooms, and cash settlements. In some cases legal costs can actually rival original construction costs.

Far beyond the direct costs, however, the indirect costs of litigation can be even more substantial. Going to court is unpredictable. Court settlements may not always reward players fairly. After all, if both sides did not believe in their own point of view, they would not be in court in the first place.

At its best, litigation seems able to mete out even, fair doses of frustration, inconvenience, and direct costs. The legal system produces compromises. It is not equipped to bring people together to devise creative solutions to their conflicts.

Perhaps even worse than any immediate conflict, using the legal system breeds a climate of blame and mistrust. People working on a project divert valuable energy to documenting their actions and defending their decisions rather than putting their heads together to work on mutual concerns and opportunities. "Standard practice around here was to open two files whenever we started a project," one construction manager commented. "One file was to keep track of project expenses, the other was to document all our actions in case we landed in court."

Partnering does not eliminate litigation but it can significantly reduce its frequency, extent, and impact for four reasons.

1. *Partnering's scheduled workshops provide an opportunity to resolve conflicts while the conflicts are small.* A partnering workshop participant commented, "Most big problems in this business start out small. It's the waiting to address them that makes them big." Because so many of the costs of a project are driven by the schedule, delays ripple throughout the project and multiply costs with each new day.

When partnering workshops are prescheduled over the life of a project, they provide checkpoints and forums to address conflicts while also training a group size that is still small enough to enhance participants' ability to solve real problems.

2. *Partnering provides face-to-face contact among the owners and senior managers of the firms involved in a project.* Without partnering this kind of face-to-face contact is unlikely to occur; there is seldom a specific reason for firm owners and senior managers to meet. When people have not met and do not know each other it is more difficult for them to attempt to resolve problems with a phone call.

Although firm owners and senior managers may not be active on a project, they play an important role in conflict resolution because they have the authority to settle problems that escalate. Once owners and senior managers meet, it becomes a little easier for projects to stay out of court because it is a little easier for them to pick up the phone and call one another.

3. *Partnering models taking responsibility for resolving conflicts.* People go to court to have someone else resolve their conflicts. This is regrettable but understandable in design and construction projects that do not have partnering because project team members are not provided with the time, place, and tools to resolve their own conflicts. Thus even if project managers advocate that people resolve their own conflicts, the typical project models and creates the conditions for a "hands-off" approach to conflict resolution.

Every component of partnering, on the other hand, brings the work of improving project communications back to the project team. Team members are the ones who write the *goals statement,* the *issue resolution process,* and the *communications procedures.*

When there is a conflict, the facilitator does not supply the solution but creates the structure and conditions so that project team members can resolve the problem on their own.

4. *Partnering sets a tone.* Say you have communications problems on two jobs. On one, you have participated in a partnering workshop with the other project team members, written a goals statement with them, compiled communications procedures, and devised conflict resolution steps. You have learned about their personality types and practiced resolving conflicts with them.

On the other project, you have not met the members of the project team except in weekly project meetings and in passing on the job site. The meetings help you get the information you need but you never really make contact with people outside your trade— everyone is in a hurry. You recognize faces and names but cannot accurately match them.

Which project has the conflict that is going to be more difficult to resolve out of court?

Partnering sets a tone that makes it easier to stay out of court. It provides contacts and interaction but also creates both explicit and implicit expectations that people will make an effort to get along with one another. You can go to court, but you are negating the efforts you yourself made in partnering workshops to get along with the other team members.

Beyond all these specific outcomes of partnering for reducing the extent of litigation, we have also encountered several examples of partnering that reduce the impact of litigation (see Table 4.1). A workshop participant recently told us that his experience with partnering was very successful even though the project he was working on ended up in court.

> We ended in court because that's the way the project was written. There was really nothing our partnering could do about that.
> What our partnering work accomplished, however, was outstanding when we finally went to court. I have never participated in such civil litigation in my life. All the sides agreed that the whole legal process worked much better because of the work we had done in partnering. The case ended up as one of the only ones I've ever heard of where everyone was satisfied with the results in the end.

Table 4.1 How Partnering Reduces Litigation Exposure and Impact

Typical project without partnering	Partnering impacts
No easy opportunity is provided to resolve conflicts quickly, easily. Small conflicts snowball.	Partnering workshops provide forum, structure, and skills to resolve conflicts quickly.
People who must work together have limited vehicles to get to know each other so they can build trust.	Partnering workshops enable participants to connect at a deeper level. If conflicts arise, it is easier to pick up the phone and discuss the issues.
There are limited means for project team members to solve their own conflicts.	Partnering models individuals taking responsibility for resolving their own conflicts.
Everyday tone of communications evolves at random, takes on a life of its own.	Partnering workshops provide a vehicle to manage the tone of communications.

Improving Project Quality. Partnering enhances project quality in three different ways: it establishes a forum in which people in different trades can discuss problems that cut across trade lines; it provides a forum in which project managers can set a tone for the project by expressing an interest in quality; it sets a model that people on the job site can emulate.

Because the work of design and construction is so fragmented among different firms and organizations, it is difficult to maintain the most minimum standards of cross-organization communications, let alone attempt to bring people together to address quality issues. Like scheduling issues, quality concerns often bridge across the lines of individual firms and organizations, so partnering provides a forum and some useful skills for the project team to tackle quality problems.

Partnering also engages the principals of the firms in dialog in which they can set a tone for what happens on the job site. Frequently at partnering workshops, project team members ask the client directly, "When we have to choose between quality and cost, which way do you want to go?" The answer, of course, is seldom clear-cut, but the ensuing discussion provides an opportunity for everyone on the project to voice their concerns about quality.

The dialog may result in some direct decisions in which quality is handled more thoughtfully. More indirectly but perhaps more important, the dialog also models a concern for quality that shapes concerns on the job site. After the partnering workshop, people on the job site inevitably ask participants, "What did you talk about?" When the answer is "Quality," people on the job site approach their work with a bit more care.

Safety. There are seldom any safety problems on construction projects, but when safety problems happen, they are extremely costly and difficult. Partnering contributes to improved project safety both directly and indirectly in much the same way that it contributes to project quality and scheduling. Like quality and scheduling problems, safety problems often arise because of the information gaps among the different trades and professions on a project.

For example, a carpenter goes to work on a wall not knowing that the electricians have just wired it. Since partnering provides a forum, perhaps the only one, where the trades and professions meet to work in a cross-project, solutions-oriented setting, it enables project team members to tackle safety problems effectively as they arise. As with quality and scheduling, partnering also provides for safety the opportunity for the project team to look ahead.

Partnering also provides a public forum to scrutinize and monitor the project, making it more difficult for people working on the project to relax their safety standards. They know that whatever they do, it's going to be discussed at the next partnering workshop. We call this partnering's "spotlight effect." Whether they themselves attend partnering workshops or not, everyone on a job that has partnering knows that the job is under a type of spotlight.

The spotlight affects the people who participate in partnering workshops; they are less likely to trim safety standards or downplay safety standards. The spotlight also affects people on the job. Knowing that their bosses and employers have attended partnering workshops, it is more difficult for them to "forget" to cross

trade lines to communicate safety-related information or point out possible problems.
See Table 4.2 for a list of partnering impacts and their sources.

Partnering's More Subtle Benefits

In addition to the immediate benefits partnering yields for projects, it also can produce more indirect, though equally important benefit outcomes (see Table 4.3).

Hawthorne Effect. When the efficiency experts in the classic Hawthorne studies tried to increase worker productivity by increasing lighting, they found that either increasing or decreasing the lighting yielded positive results. Ultimately they realized that workers improved performance more as a result of the researchers' attention than of the increased wattage. They labeled this unexpected positive outcome, resulting mostly from paying attention to the workers, the "Hawthorne effect."

Similar results occur in partnering. Designating a project as a partnering effort means that the project receives extra attention. Architecture, engineering, contractor, and subcontractor firm owners watch the project a little more closely and oversee their people on site a little more carefully. The increased attention increases the care with which people communicate and this in itself reduces conflict.

Halo Effect. Simply attaching the label "partnering" creates expectations for how people on the project ought to communicate with one another. People who attend partnering workshops return to the site attempting to carry the spirit of partnering into areas beyond those covered in the workshop agendas. Even people who do not attend workshops take extra care in communicating after they see the partnering posters and memos on the site.

Spotlight Effect. Just as partnering tends to increase the overall level of attention people pay to communications, it often also helps reduce the more outrageous kinds of communications that can lead to conflict and litigation. When tempers explode on a construction site that has partnering signs on the walls of the con-

Table 4.2 Project Impacts of Partnering and Their Sources

Construction project impact	Sources in partnering program
▪ Reduced costs resulting from reduced time lines and improved schedule monitoring	▪ More frequent discussion of schedules and easier opportunities to make adjustments when problems are discovered ▪ Opportunity to work as a group to discover ways to trim schedules
▪ Reduced extent and frequency of litigation	▪ More frequent opportunities to discuss conflicts. ▪ Opportunity to discuss conflicts while they are small and more easily resolvable ▪ Partnering provides face-to-face contact and makes it easier to pick up the phone later ▪ Partnering encourages and models taking individual responsibility for conflict resolution ▪ Partnering sets a positive tone
▪ Increased building quality	▪ Workshops provide opportunity to address quality problems and explore opportunities
▪ Improved safety	▪ More frequent discussions of project, both to address specific issues and to plan ▪ "Spotlight effect" that the project is being monitored and observed by more people increases individual mindfulness
▪ Overall	▪ Partnering provides (sometimes the only) forum to tackle job problems such as costs, litigation, quality, and safety that cross trade and profession boundaries

tractor's trailer, someone is usually quick to remind the combatants that "You can act like that, but we're going to discuss it at the next partnering workshop."

Even if the people involved in the conflict cannot immediately move to a more collaborative approach, the knowledge that the

Table 4.3   Unanticipated Benefit Outcomes of Partnering

Outcome	Specifics
Hawthorne effect	People work harder at communications because they know the project is receiving attention
Halo effect	People outside the immediate circle of partnering workshops improve communications since the project is "supposed to be partnering"
Spotlight effect	People who might act outrageously temper their behaviors because they know the project is under the spotlight of partnering
Equalizer effect	Partnering levels project hierarchies by putting people to work together on cross-level teams
Accountability effect	Because they know it will be discussed, people become more accountable for their behavior
Internal communications effect	Organizations that participate in partnering often learn about internal problems in processing information and communications
Marketing effect	People make new contacts in a meaningful way in partnering workshops
Management and professional development	Participants acquire management and professional development skills and concepts

project is under a spotlight helps them curb kinds of actions that could get them into trouble later.

Equalizer Effect. Many job sites have an unstated though distinct hierarchy or pecking order. Craftspeople talk to each other but not to supervisors and even less to engineers and architects. This can result in serious information gaps between the levels of the hierarchy.

Partnering does not eliminate the hierarchy but it does create spaces and places during the project in which the playing field is leveled. Working in a small group on *goals statements, communications procedures,* or *conflict resolution* places people on equal footing in the workshop. In addition, the connections people make in the workshop help them to make necessary contact back on the job site with less hesitation and greater comfort.

Accountability Effect.   Most people associate partnering with reduced conflict and improved communications, neglecting the "harder side" of partnering: increased accountability.

By bringing communications among team members into an open forum, partnering increases individual responsibility and accountability. When project team members discuss their complaints with one another and devise resolutions in partnering workshops, the whole project team is often listening. It is very difficult for project team members to "forget" the agreements they make in a partnering workshop.

Internal Communications Effect.   Firms that participate in partnering workshops often discover that the cause of a communications problem or conflict is their own internal communications. For example, a project architect submits a change order in response to a pressing problem raised by a contractor, but the architect's firm takes two weeks to process the order. The delay results in cost charges for the contractor, and conflict between the contractor and the architect on site. The contractor accuses the architect of not following through on the change order but the real culprit is the architect's firm, which was too slow. As a result of the problem, the architect sets about reviewing how the firm works internally and forges improvements that result in improved processes and procedures.

Marketing Effect.   New and expanded business relationships often come out of participating in partnering workshops. Because effective communications are so important to the success of a project, people are reluctant to choose to work with people they do not know. Yet there are few ways for people in design and construction to get to know one another in a meaningful way. Professional meetings usually provide an opportunity for only a brief encounter or handshake.

Partnering, however, provides an in-depth kind of interaction among workshop participants. People who work together in a partnering workshop spend a full day or more working in small groups on problems of mutual interest. There are few more effective vehicles than this for people in the design and construction field to get to meet new people and to learn a great deal, first hand, about how they work together.

Management and Professional Development Benefit. For decades now, many contractors, and quite a few architects and engineers as well, have listened in silent envy while their friends and neighbors who work in other kinds of businesses have described their experiences with professional development and training. Although people in the design and construction industry advocate on behalf of training, they participate less often than do their peers in other businesses.

For many, partnering provides the first substantive professional development activity since school. Partnering provides participants with many of the same skills and concepts they would encounter in the best management and professional development programs in other businesses. Moreover, participants often acquire skills they can put to practical use in other aspects of their business.

Benefits to Specific Project Team Members

Members of a typical project team bring to partnering different concerns arising from their traditional, specific tasks and perspectives. They benefit from participating in partnering in two ways: first in the traditional, established concerns of their respective trades and professions, and second, through unanticipated benefits and impacts. The second, unanticipated outcomes of partnering result from the new information partnering brings them along with the new ability to act on that information (see Table 4.4).

Architects. Architects we have worked with usually have two different reactions to partnering. Their initial reaction is "This is nothing new, we've done this before, back in the 1960s when we called it 'participative design,'" or "This is pretty much what we do with our clients already." Once they see what partnering really is, though, they often back off and admit that it is much more organized, focused, and potentially much more effective in accomplishing the goals of client involvement than their more informal efforts attempt.

Once they realize how effective partnering can be, architects often back off. "I don't know," one hesitates, "I like client involve-

Table 4.4 Benefits of Partnering for Specific Project Team Members

Stakeholder	Immediate benefits	Long-term benefits
Architect	Provides structured dialog to protect the integrity of the design	Provides structure and skills for managing project communications
Owner	Provides a tool to obtain more timely information on project	Provides easy forum for expressing preferences and opinions about project
Contractor	Addresses and improves immediate costs of communications delays	Places contractor in more positive, problem-solving role in the project team
"Minor players" (subcontractors, specific engineers, consultants, etc.)	Provides them with more complete information on the project	Enhances their ability to communicate so that they are heard; overall, "levels the playing field"

ment and all that but this looks like it could be too much. It looks like you could lose the whole design here, give it over to the client." Architects often worry about protecting the integrity of their design. They express concern that partnering opens the design up to so much participation from the client as well as from contractors and engineers that they will have to devote a great deal of effort to defending it instead of simply carrying it out.

Architects have been pleased with the results of partnering at two levels. First, they find that it helps them with their own initial concerns of protecting the integrity of the design. One architect told us

I was worried that opening up my design to discussion and input from all the subcontractors would dilute the statement I was trying to make. What actually happened, though, was that partnering provided a device for organizing all the input that comes up on a project anyway. I realized that [in] the projects I had trouble with, the trouble came about because there was no forum like partnering to field and manage the input from all the members of the project team.

Architects also find that partnering helps them improve their effectiveness in managing communications for the project. A designer explained.

> We spend all our time in school on design, then when we build something we find that we're spending most of our time producing working drawings and trying to manage communications on the project. Needless to say, there aren't many of us who are as good at managing design communications as we are at design.
>
> Partnering has helped me with the design communications tasks. I found out exactly what people need in the workshops, and the workshops have helped me develop better methods to manage the flood of communications that goes with every project. Partnering doesn't make the unpleasant tasks go away, but it helps me deal with them more effectively.

Owners and Clients.   Clients often struggle with participation in projects. They need to participate in some way so that they get the building they want, but they often do not get the information they need to exert influence effectively. Project meetings provide them with some useful information and provide an opportunity for input, but a great deal can happen between meetings.

Other members of the project team often come to view the client as "the enemy" and with good reason. Clients make inconvenient demands, change their minds, create delays, and cause cost overruns they don't want to pay for. Thus clients often get involved in conflicts with other members of the project team and without partnering, they seldom have a forum or a set of strategies and skills to resolve their conflicts.

For clients, partnering provides one more vehicle to get information about the project and one more means to exert influence over the project. When the client is a large organization or government agency, partnering provides the additional benefit of establishing goals and procedures that contribute to continuity on the project even if the client organization experiences turnover, layoffs, or reorganizations.

A state agency director explained, "The project we had last year that used partnering was the first one in my experience of 22 years where I actually felt like I had a voice in a useful way. Sure, I participate on all my projects but I usually have to fight to have my say. With partnering we all could put our efforts into solving problems instead of having to convince others that a problem existed."

Contractor. We have been surprised that in most of the partnering workshops and projects we have conducted, the contractors have been the most enthusiastic supporters of partnering. When we have asked the contractors about this, their comments have resembled those of a contractor from the midwest,

> Of course we're interested in partnering; we're the ones whose reputations are on the line in dollars and cents. The architects' work is already done, so we're the ones who lose the money if we can't keep building. We're the ones who have to pay the plumbers overtime if there's no work for them to do.
> Also, we're the ones who are usually in the position of looking for information. The architects have it; they're the ones who can tell us what we need.
> Because of the way the whole thing is set up, we're always going to be interested in anything that gets us the information we need, and that's where partnering comes in. It isn't perfect, but it's better than anything else I've seen.

The secondary benefit of partnering for many contractors is that it takes them out of the role of "complainer." "For years now, I've been thinking of myself as 'the nag,'" a west coast contractor explained.

> When we started participating in partnering I started realizing that it's not me. I can admit that there are legitimate problems, I'm not an obsessive person. I look like the problem and the architect looks like the good guy.
> Partnering puts me in a more positive light. In real time, I spend less of my days trying to track people and information down, less of my time on the attack. I spend more of my time on problem solving.

"Minor Players": Engineers and Subcontractors. Architect, client, and contractor dominate most of the discussion on design and construction projects, but a large number of other players participate on the project team: engineers, subcontractors, and consultants. These people perform specialized tasks and usually work just on one phase of the project. They have special information and participation needs that partnering addresses.

The minor players need to get the information they need in order to do their job, even though they are not involved over the full course of the project. They also need to exert influence over the other project team members such that the project team members use the information they bring to the table. Yet because they are involved in just a brief time period or scope of project work, they are seldom able either to get the information they need or to influence others effectively with regard to what they bring to the table.

Partnering "levels the playing field" for the more minor members of the project team. The structure of partnering makes it easier for them both to get the information they need and to present their point of view in such a way that others can understand it and use it.

A Manhattan structural engineer summed up partnering's effects for him:

> On my projects where we don't have partnering, I usually see the same thing at project meetings. Remember those old commercials—"When E.F. Hutton talks, people listen"—where everybody drops what they're doing to hear the one person who has information? It's like that for me, but in reverse. When I talk, people ignore me. They take out other reports, they look at their calendars, sometimes they even read the newspaper.
>
> I can understand it. Most of the time, they really don't need to know much about what I'm saying and recommending. In the rare cases where they really do need to hear me, unfortunately, they don't know enough about what I'm saying to use it well.
>
> Partnering, on the other hand, makes my job a little easier. The workshops give me a forum in which people pay more attention to what I have to say. They listen; they often actually discuss it. Later on, if they need to use it, they can.

5

Deeper Impacts: What Partnering Does and Why

Summary

The benefits of partnering described in Chapter 4 are exciting and important. The deeper potential impacts of partnering described in this chapter are compelling. Beyond the immediate project benefits described in the previous chapter, partnering can also result in deeper impacts on some of the defining conditions and most fundamental aspects of design and construction work.

This chapter begins to explore partnering's deeper impacts by examining in greater detail the highly fragmented design and construction industry. The chapter then illustrates partnering's deeper impacts in three areas:

- The nature of individual project team member participation and involvement on projects
- The design products resulting from partnering
- The organizational issues of firms in the design and construction business

In order to understand fully how partnering can create these impacts, we explore in more detail what partnering does to alter

and improve the nature of communications on a project and conclude with a summary list of reasons why partnering works the
way it does.

Babel: Conditions in the Design and Construction Industry

With its tale of different people needing to communicate but
speaking different languages, the Tower Of Babel story paints a
graphic picture of some of the coordination and communications
problems built into the design and construction industry. The
highly fragmented nature of the design and construction industry is one of the factors that shapes and defines every project.
These problems create the context in which partnering operates.

There is some fragmentation in all large organizations. Every
day, products are delayed because sales departments don't talk to
manufacturing departments. There's trouble in many companies
with quality because manufacturing groups don't talk to engineering groups. Customers of many businesses jump ship
because product design teams do not listen to customer service
departments. The Internal Revenue Service audits companies
because nobody listens to accounting departments.

Organizations with all operations under one roof encounter serious interdepartmental communications gaps. With its divisions into
many organizations under many roofs, it's no wonder that design
and construction encounters serious communications problems.

More than other businesses, design and construction are highly
fragmented. The typical manufacturer of the 1990s works to
reengineer internal processes, enlarge job definitions and responsibilities, reduce cycle times, and build cross-functional teams to
integrate operations. Without partnering, the typical design and
construction project of the 1990s splits hairs over which trade is
not responsible for what problem.

In design and construction, much more attention has been paid
to dividing up the work of making a building than to putting all
the pieces back together again. Design and construction comprise
an industry of specialists. Engineers don't design, architects don't
engineer, carpenters don't do plumbing. The many types of engineers (civil, structural, plumbing, electrical, roofing, telecommunications, etc.) ply their own professions with little concern over

how to interface with the others. The building trades divide and then subdivide into increasingly narrower specialties: asbestos removal, hazardous waste handling, site safety coordination, sheetrock installing, suspended ceiling construction, etc.

On the positive side, this extreme specialization makes it possible for individuals and firms to refine skills in their areas of expertise. It would be difficult for any one person to be an outstanding structural engineer, designer, and plumber. Specialization also creates some short-term efficiencies. It would be difficult for the same firm to learn to become profitable at both conceptual design and building drywall.

Job specialization also means that each trade and profession can address issues within its own scope with a high degree of skill, experience, and success. It is rare, even on a complex job, to hear of a delay caused by one of the trades being unable to figure out how to carry out its job.

While job specialization creates some benefits, it also creates some problems. Specifically, any of the project tasks and issues that involve interaction, coordination, or interdependency among the trades and professions are often problem spots: scheduling, costs, quality, and planning.

Whenever a project experiences problems in any of these areas, each of the separate professions and trades usually works hard to defend its innocence and blamelessness, often working with an equal or greater amount of effort to attempt to pin the blame on another group. In the end, it is the project that suffers most as a result of all this, as little attention and energy is devoted to actually solving the problem.

These areas of project coordination and interrelationships form the basis for much of what partnering attempts to address. Partnering accomplishes most of its work not within any of the professions and trades but in the spaces left between them.

Why Task Fragmentation in Design and Construction?

Task fragmentation in design and construction is so deeply rooted in tradition and culture that it is sometimes difficult to discern why it is so extensive (see Table 5.1). A number of different factors contribute to task fragmentation in the industry.

Table 5.1 Task Specialization in Design and
Construction

Benefits	Problem areas	Roots

▪ Specialization
▪ High levels of performance within trades and professions
▪ Efficiencies
▪ Technical expertise
▪ Professional identity, camaraderie
	▪ Any task involving interdependencies, interaction and communications across trade and professional lines:
	  Scheduling
	  Costs
	  Quality
	  Planning
		▪ History, culture
		▪ Different loyalties
		▪ Tunnel vision
		▪ Different languages
		▪ Professional mythologies, stereotypes, and jealousies

1. *A History of Its Own.*   Unlike software, telecommunications, electronics, aircraft manufacture, or many other sizable parts of the economy, design and construction together are one of the oldest existing industries. They have a combined history and strong culture all their own, dating back thousands of years to the pyramids and earlier. Many of the job and task divisions people take for granted now are rooted in a history no one can remember let alone explain.

This extensive history has left a legacy of "standards," widely used practices and understandings that add up to make any kind of change difficult. Any kind of change in the industry runs headlong into thousands of years of "We never did it that way before."

2. *Fragmented Loyalties and Incentives.*   When plumbers on a job site see a carpentry problem, it is not always in their interest to bring that problem to anyone's attention. Addressing the problem might impact overall project scheduling, costing them time and money. Even if there is no notable cost inhibiting the plumbers, there is rarely an incentive for them to report the carpentry problem. In fact, there are few incentives and often some costs for communicating outside one's immediate discipline.

3. *Tunnel Vision.*   Plumbers on the job site who fail to report the carpentry problem are also likely to be driven by the fact that they don't see the carpentry problem in the first place. With years of work rules' experience restricting their view, plumbers focus on

the task at hand. They can easily, simply miss even glaring problems created in other trades and professions working around them.

4. *Professional Myths, Jealousies, and Stereotypes.* The work of design and construction has been divided long enough for the various professions and trades to have developed their own mythologies. These can be entertaining in some ways, as in some of the bumper stickers the different trades sport: "Carpenters do it on the level," "Electricians do it with energy," etc.

Professional mythologies also produce professional jealousies and stereotypes, and these help to keep the trades and professions apart. When architects extol that "All great architecture leaks," for example, the subcontractors hear it as arrogance. They knowingly shake their heads about "those people" and prepare to do battle over requests that the individual architect on the job might actually find useful.

5. *Different Languages.* Should people on the project team attempt to communicate across the boundaries of their trades and professions, they are likely to be frustrated by the different languages they speak. Architecture attracts a different mix of personality types than does engineering or contracting. When they all get together, the engineers and contractors often complain about the architects' lack of specificity and detail in the drawings (particularly 3¼ inches down on page 6). The architects often respond with an analogy: the drawings are like a road map, a cloud, or a song. As the discussion continues, both sides press on and diverge more and more.

Deeper Impacts: Stages of Project Participation and Involvement

Partnering impacts the fragmentation of the design and construction industry by enhancing project team members' ability to break through the the traditional boundaries of their trades or professions and participate more fully in the project (see Fig. 5.1).

Stage 1. *Not my job.* This is the most powerful result of task fragmentation. Project team members reason, "I participate in my own trade only. Any other consideration is 'not my job.'"

Stage 5. "Synergy"

Team members use individual differences constructively. Initiate project-wide perspective with expectation of improvement.

Stage 4. "Team Player."

Project-wide scope of interest. Attempts to cooperate with other trades and professions

Stage 3. "Use my resources."

Willing to participate outside traditional scope but only if asked.

Stage 2. "How does this affect me?"

Interested, but only in narrow, self-oriented focus.

Stage 1. "Not my job."

Disinterested, apathetic.

Figure 5.1. Stages of involvement and participation in a design and construction project.

Stage 2. *How does this affect me?* Task fragmentation may also lead to this narrow focus. "I participate in my own trade, and I'm interested in some other aspects of the job, but only the ones that affect how I perform my trade."

Stage 3. *Use my resources effectively.* When trade and professional boundaries begin to soften, people begin to think in the following way: "I participate in my own trade, any project-wide information relevant to my trade, and in the ways my trade connects with the overall project." Here I am concerned not only with the information I take in, I also want to get information out about connecting my trade to the rest of the project.

Stage 4. *Team player and project-wide interests.* Here people break traditional identities and limitations more, taking the stance, "I'm interested in the project overall, in things that impact the project, in how the project is going. I may be able to make immediate use of this information or impact how my trade is used but even if I can't I want to know because I look at myself as a member of the project team, not just as a practitioner of a particular trade."

Stage 5. *Synergy and idea development.* This is what is possible when people become even more identified with the project instead of with only their own trade or profession. "I try to take the ideas and perspectives others on the project provide and play with them, run them through my own frame of reference. So doing, I am likely to see their points of view differently and thus use them to mutual advantage in the project. For example, I ask the plumber to comment on my drawings. Instead of defending my point of view, I work with what the plumber offers. At first I usually disagree with him but if I can discipline myself to work with his ideas, not just react to them, I can usually develop them in a way that he didn't think of himself."

The overall project benefits through increased idea development. The heating, ventilation, and air conditioning (HVAC) contractor develops the plumber's ideas in new ways, and then hands back to the plumber a developed idea the plumber could not have produced on his or her own. The overall effect of this approach to idea development is like two members of a football team lateralling off to each other whenever each encounters an obstacle on the field. The ball moves downfield because of both of them. The ball moves because of their individual efforts and because of their skill in seeing and responding to each other's opportunities and problems.

Partnering does not guarantee stage 5 participation but it creates conditions to support it. When the project team gets together to look into an unexpected delay, for example, it is in everyone's interest to develop ideas for a solution. The engineer may shed new light on an architectural problem, the contractor may offer new insight into an engineering problem.

Partnering does more than create the conditions for stage 5 participation; it creates some expectations that such idea development will occur. Simply bringing people together to address mutual problems and providing them with some basic training in creative problem solving creates clear expectations and hopes for how the group is to proceed. At such a meeting it would be difficult for the HVAC contractor to participate in a discussion aimed at devising new solutions to a project delay by saying, "It's not my job" or "How does this impact me?"

Deeper Impacts: Individual Project Participation

To understand more clearly how partnering impacts the fragmented nature of design and construction work and the ways in which project team members can participate, we examine two different dimensions of project team members' participation: passive to active and specific to universal ranges (see Table 5.2).

Passive to Active Range.   In ineffective teams, project team members are passive, waiting to be told what to do and limiting their actions to short-term, unthinking fulfillment of orders. When team members are more active in the process, they bring more information to it. Fragmented work encourages a more passive stance towards the overall project because team members neither get enough information to participate more actively nor are they provided with any means to act on project-wide interests.

Partnering may not "make" project team members more active in project communications, but it provides opportunities to be

Table 5.2   Participation on the Project Team: Scope by Activity Level

	Scope of Participation	
Activity level	Universal	Specific
Active	*Project-oriented, collaborative:* HVAC contractor works with electrical contractor to devise new schedule that works better for both, and for overall project as well.	*Self-oriented, defensive, protective:* HVAC contractor wants to do his or her work before the electrical contractor in spite of overall higher costs to the project.
Passive	*"Interested":* HVAC contractor is interested in the overall project but has no vehicle to express that interest.	*Apathetic:* HVAC contractor is interested only in finishing HVAC work, not in overall project.

active that are not available in any other way. With all these activities, partnering makes it more difficult for a project team member to remain passive:

- At minimum, people attend workshops.

- The workshops provide a place, a forum where it is easier for people to discuss their concerns.

- Partnering provides effective vehicles for discussion.

- Partnering provides information on the project and opportunities to comment on it.

- Partnering opens up the problem-solving process of a project, making it easier for more team members to participate.

Specific to Universal Range.   Being active in itself is not enough to improve the quality of design, though. It is also necessary that when active, team members focus their efforts on concerns that affect the project overall. If they focus stubbornly and narrowly on their own concerns (as fragmented work would encourage them to), project team members will likely detract from the design.

Partnering increases the scope of project team members' concerns from specific to more general by:

- Having them work with conflict resolution skills that model concern for and involvement with the other person's position and ideas
- Exposing them to others' concerns in workshops
- Putting them together with people who have different concerns to collaborate in producing mutual tasks such as goals and issue resolution documents
- Creating time and space in workshops to use ideas brought to the design specifically by others

When project team members are both active in the process and universal in their focus, they bring more information to the design. More information circulating in active discussion and dialog makes it more possible to generate more ideas about the overall project. More people bring richer ideas to their dialog with the architect.

Deeper Impacts: The Design Product

Beyond its deeper impacts on the design and construction process and on the scope of involvement of project team members on the project, partnering also impacts the product of the design process by bringing more thinking, idea development and problem-solving effort to the project.

1. *Partnering helps project team members protect original design intent.* Any experienced architect knows that even after they are approved by the client, the best designs can be significantly revised and seriously compromised during actual construction. Often it isn't until the building is being built that it becomes clear that a key aspect of the design is not feasible. At that point if there is not an effective means of communication between the architect and the project team, the original design intent is all too likely to be lost.

2. *Partnering brings project team members more ideas, information, and data.* The increased communications resulting from partner-

ing during construction brings more ideas, information, and data to the designer. This increased amount of information can lead to better design as the designer can more effectively respond to the situation and other team members.

The extent of this impact depends in large part on how early partnering is incorporated into the project. The sooner partnering is used, the more information it brings the designer. Information brought in after construction begins has much less impact on the design, as it is usually too late to allow it to influence the design approach.

Viewing partnering as a potential resource to bring new and useful information into the design process makes it clear that the value of partnering is increased if it is begun as early in the process as possible, well before construction begins.

3. *Partnering models a more effective, reflective inner dialog for all project team members.* People involved with partnering encounter structures that encourage the continual discussion of ideas and issues. Ongoing partnering workshops force not only a discussion but a continuation and expansion of the thinking process. In collaborative discussions, the design ideas are discussed, reviewed, examined, and explored. The simple, traditional design model of think–decide–execute is replaced with the more complex model of think–decide–discuss–review.

Potentially, architects who encounter the more open approach to design in partnering discussions will bring an increased amount of inquiry and exploration back to their individual approach to design.

4. *Increased communication increases the amount of thinking others do about design issues.* Knowing that they will meet every few months to discuss the project, clients, engineers, contractors, and others involved in partnering increase the scope of their own thinking about the project. At the outset, people involved with partnering tend to focus on protecting their own turf. Over time, however, as they learn conflict resolution skills and get together to discuss the project overall, they move more to think about a wider scope of issues. They come to workshops not only with complaints but with possible resolutions to those complaints.

Overall, then, partnering increases the total amount of thinking being done about the project. The process produces more ideas not only for the architect but for all the other members of the design process. They have more information and they have a forum in which to develop that information.

Deeper Impacts: Organizations

In addition to its impact on participation in the design and construction process, partnering also exerts deeper impact on the organizations that participate in it. All businesses face difficult economic and financial challenges in the economy of this era, but the challenges that organizations in the design and construction business face are particularly difficult. Partnering helps the organizations that participate in it (both service providers and client organizations) to respond more effectively to the business challenges they confront.

Conditions.   The project-based nature of design and construction creates an inherent, underlying condition of instability that poses a major organizational challenge to service provider organizations as well as clients.

From the perspective of architecture, engineering, and construction firms, work flow is unpredictable and unsteady. Every business faces some instability in this regard, but not as much as design and construction where every new job must be won and successful job completion means moving on to a new project. Banking, manufacturing, retail, and even farming do not face this kind of instability and unpredictability.

The project-based nature of design and construction favors organizations that can respond quickly and efficiently to client requests for proposals and demands, form teams rapidly to get a job done, deliver a product on time and then walk away quickly, re-form, and move on to the next project. Manufacturing organizations, aware of the value in this kind of quick change, are learning to reorganize and refocus as quickly and effectively as design and construction firms can.

Problems.   On the other hand, design and construction firms' ability to re-form and keep a sharp project focus often leads to chaos and ambiguity of internal conditions. While firms focus on projects, they neglect their own internal infrastructure and communications procedures. While the project teams usually enjoy effective internal communications, information exchanges between project teams often leave a great deal to be desired.

As a result of the project focus and information gaps among project teams, people working in the typical design, engineering, or construction firm often don't know what's going on with projects they're not working on directly. Thus they frequently duplicate efforts, miss opportunities given under their own roof, and fail to learn optimally from their own experience.

In the project-based firm, project management skills are required but organizational management skills are often not. Project management tools are provided but internal organizational infrastructure is often neglected. In daily life, this translates into architecture, engineering, and construction firms being the last businesses on the block to implement internal communications tools, everyday computer applications (beyond computer-assisted design), and accounting systems that efficiently report the financial information a firm needs.

Partnering Impacts. Partnering does not change the project-based nature of design and construction but it turns firms that participate back to attend to their internal operations and infrastructure. Firms participating in partnering frequently discover that their internal communications among project teams are not effective. They repair these systems in order to meet partnering commitments and end up with improved systems that carry over to other projects.

For example, architecture firms that participate in partnering often discover a communications gap as a result of efforts to improve the request for information process. At a partnering workshop, the project architect commits to what seems like the reasonable goal of turning the contractor's requests for information around in 24 hours or less.

When the contractor complains at the follow-up partnering workshop that requests for information are being delayed, the architect returns to the firm to discover that his or her own project team is returning requests quickly. The most serious gaps and delays are occurring mostly in cases where people working on this project need to get information from people on other projects.

The commitments the architect made in partnering encourage him or her to attend to the cross-project communications issue for the partnering project. One architect even commented, "The promises I made in the partnering workshop gave me the mandate I had been looking for for years to get people in the firm to finally pay attention to internal issues and communications."

Moreover, because the internal improvements firms make in order to keep their partnering commitments often affect systems or infrastructure, they leave lasting results in the firm. Architects, engineers, and contractors who have had to improve their firms' ability to provide information in partnering usually carry their new insights and skills on to the next project.

The Systems Perspective. Another way to understand the impacts of a project focus is with the systems perspective. Focusing on project-to-project issues moves firms towards becoming closed systems, i.e., organizations that take in less information than is optimum from their environment. Partnering makes firms into more open systems, bringing in new information.

The premise that design and construction firms are closed systems may seem illogical, as a project focus forces firms into communications with clients and with each other in order to win bids and build buildings. Of course a kind of openness does exist, or else firms would be unable to conduct business.

At the same time, however, the highly fragmented nature of the design and construction business (see earlier in this chapter) also limits the amount of information firms take in. It is perfectly possible for firms to win a bid with little exposure to potentially valuable new ideas, practices, insights, and opportunities occurring in the organizations of their project team members or in the client organization. A project focus forces interaction but it also encourages limited openness.

Partnering places members of the firms into dialog with one another in which it is natural to discuss and share information about new processes, innovations, improvements, and management practices. Partnering makes it possible to have a level of depth in exchanging ideas that does not exist in other aspects of the design and construction business except perhaps in formal training programs and seminars.

Partnering opens up the traditionally closed systems of design and construction firms with new ideas, practices, and insights that can invigorate the process of organizational learning and improvement (see Table 5.3).

Table 5.3 Partnering Impacts on Architecture, Engineering, and Construction Firms

Conditions	Problems	Partnering impacts
Project-based, project emphasis	Effective communications within project teams, gaps between teams	Partnering surfaces internal gaps that need attention
Unstable, unpredictable workload	Responsive to changes, neglect continuity, limited organizational learning	Partnering elicits commitments that drive lasting internal changes
Project management emphasis, organizational management neglected	Project resources available, firm-wide and systems infrastructure neglected	Partnering requires internal systems improvements
Emphasis on projects limits in-depth interaction with clients and other organizations	Firms become closed systems, unaware of other organizational forms and opportunities	Partnering exposes firms to other organizations in an in-depth way enabling new information to come in

Deeper Impacts: Communications

While the previous sections of this chapter examine what some of the deeper impacts of partnering are, this section examines in more depth *why* partnering produces such impacts. Specifically, this section explores how partnering alters and improves the nature of communications on a project (see Table 5.4).

1. *A forum.* Partnering provides a forum for discussion that is otherwise missing in the design and construction process. Design and construction break work down into small pieces, dividing responsibilities, tasks, and roles with rigid lines and doing nothing to make sure that necessary information flows across the boundaries.

2. *Proactive.* Acknowledging from the outset that miscommunications and conflict are inevitable, partnering works with participants to help them anticipate and then effectively manage the communications problems that arise.

Table 5.4 Partnering Program Components' Impacts on Project Communications

Program component or aspect	Impacts
Forum	Partnering provides a regularly scheduled forum for discussion otherwise missing in the design and construction process. Simply having a forum creates a place in which to discuss conflicts and miscommunications.
Proactive	By anticipating problems and planning responses to them, partnering takes a proactive approach to communications.
Specific and procedural	By creating specific communications procedures, partnering provides clear guidelines for improving communications.
Strategic, conceptual	Partnering also addresses more global concerns, involving participants in discussing their individual project goals.
Legal overtones	Program participants are asked to sign the goals and communications documents they produce, thus increasing their commitment.
Training as well as consultation	Partnering provides depth to conflict resolution by also training project team members in communications skills.
Empowerment	By dropping decision-making responsibility to the lowest levels possible, partnering also increases communications efficiency.
A synthesis	Because partnering combines and synthesizes all the above, it impacts the project at a wide variety of levels.

3. *Specific and procedural.* Unlike many impractical organizational exercises, partnering puts project participants to work devising detailed, specific procedures to maximize information flow.

4. *Strategic, conceptual.* Unlike some overly detailed organizational activities, partnering also elevates participants to discuss global concerns, personal goals, empowerment, and project possibilities.

5. *Legal.* With strong roots in an intention to reduce construction claims and litigation, partnering uses legal tactics to ensure that participants follow through on their specific and strategic commitments. Specifically, participants sign the goals and procedures they devise.

6. *Training.* Unlike some purely tactical consulting efforts, partnering also includes a significant amount of training intended to enable participants to improve their everyday interactions with one another.

7. *Empowerment.* Most partnering efforts attempt to give project team members increased autonomy and control by dropping decision-making authority down to the lowest possible levels. This improves the effectiveness of decisions because people with first-hand information on the issues deal directly with each other.

8. *A synthesis.* Nearly all other attempts to train people or develop organizations put most of their efforts into one major approach. By diversifying and using many approaches, partnering is less likely to miss an important issue.

Underlying Principles: Why Partnering Works

To conclude a deeper understanding of the impacts of partnering, we explore in greater detail why partnering works and discover a number of different reasons with different sources (see Fig. 5.2):

1. *Communications effectiveness significantly impacts overall project success.* Project communications contributes significantly to overall project success. While other aspects of project performance and productivity have been relatively easy targets for technological and systems improvements, communications have remained problematic. At the same time, communications problems have become increasingly troublesome sources of cost overruns, delays, and litigation. Anything that offers some kind of improvement in this troubled area will capture peoples' attention and effort.

2. *Communications can be managed.* Project communications are manageable in two ways:

---

1. Communications significantly impacts overall project success.
2. Communications can be managed:
   - Predictable problems.
   - Systems solutions.
3. Relationships need maintenance.
4. In communications, proactive works better than reactive.
5. Partnering convenes key decision makers.
6. Team building methods have evolved.
7. Groups can govern themselves if they are structured and organized.
8. Partnering can provide the skills training people need to improve project communications.
9. Partnering involves a strategic combination of strategies and skills.
10. Partnering emphasizes follow-through activities.

Figure 5.2. Why partnering works.

- *Predictable problems.* Similar communications problems occur on most projects. (For example, contractors frequently complain to architects that their working drawings do not contain adequate detail.) This predictability makes it easier to anticipate what may go wrong and take corrective action.
- *Systems responses.* Many of the predictable communications problems of a project can be remedied with improved systems for managing communications. For example, keeping an accurate and timely log of RFIs (contractors' request for information made to the architect) helps manage the conflicts that arise when contractors complain that architects take too long to respond to their requests.

3. *Relationships need maintenance.* Married couples typically set aside time, often regularly and frequently, to unravel differences, review and resolve standing problems, get caught up with each other, and provide encouragement, support, and reassurance. These same kinds of maintenance activities are necessary in

any working relationship, and they are especially important in partnering.

4. *In managing communications, proactive works better than reactive.* In one of the first partnering workshops we conducted, we were beginning to work on conflict resolution processes, and the group was holding back. The general contractor intervened, addressing the group, "Haven't you ever worked on a project before? You know you're going to argue, you know you're going to have disagreements. If you know this, then why don't we plan how to handle it now, when we're calm and reasonable, instead of waiting until we're fighting?"

Working on conflict resolution processes, a project team is usually able to predict which conflicts will occur, devise means to resolve specific conflicts and design more general systems that function effectively to make sure information is exchanged effectively and conflict never arises.

5. *Partnering convenes key decision makers.* When project team members work on resolving a particularly thorny conflict they often achieve success fairly quickly. When this occurs, one of them usually comments, "A big part of the problem was that we never could hear from everyone involved at the same time."

In everyday terms, this is evident when project team members can't find the right person to make a decision or supply an opinion. This also happens when the people working on a project can't make a decision on their own, and when the people who can make the decision are on vacation, traveling to a conference, or unresponsive to calls and memos. Partnering cuts through internal communications by bringing the key decision makers into the same room at the same time.

6. *Team-building methods have evolved.* Outside design and construction (often in client organizations), organizations currently make extensive use of team-building methods and strategies that resemble partnering. In fact, partnering is really a form of team building applied to design and construction projects. Team building has evolved over the years beyond fad or "program" to become, for many companies, an integral part of the way the company does business.

The result of this evolution for design and construction is as follows:

- *Proven methods.* Partnering strategies and skills have an extensive track record as team-building strategies and skills in other organizations.
- *Experienced participants.* Often people participating in partnering for the first time bring relevant experience from team-building workshops they have participated in in other organizational settings. Often, for example, they are familiar with writing and using a goals statement. When people have this kind of experience they can usually get more from partnering and help partnering work more effectively for the whole project team.

7. *Groups can govern themselves if they are organized and structured.* When people who have worked only in hierarchical organizations encounter the idea in partnering that the project team will make its own decisions to resolve conflicts or manage communications, they frequently express skepticism.

"All this is fine," they typically comment, "but we all know that at some point, you're going to need one person to take charge."

That may be true, but the more skilled the group and facilitator are, the more decisions a group can make on its own. For example, a project group may split into two camps about the best steps to follow in conflict resolution. A skilled group will step back from the argument at hand and ask members, "What are several different ways we can handle this split?" With the conflict reframed in this way, the group can usually decide.

The team-building literature makes it clear that groups can learn to govern themselves if they acquire some critical communications skills, work with a skilled facilitator, and follow through on their promises and commitments.

8. *Partnering can provide the skills training people need to improve project communications.* Although effective communications account for a major component of project success, most design and construction professionals and craftspeople resemble managers and supervisors in manufacturing, research, retail, computers, and most businesses. Their technical skills in their field far surpass their skills in communications.

Partnering places a great deal of emphasis on creating systems for effective communications, and many of those systems do not depend for their success on project team members' individual communications skills. However, much of project team members' ability to implement partnering agreements in goals statements and conflict resolution processes depends on their ability to listen and resolve their own conflicts with some comfort and skill.

Partnering often provides (we believe it always should provide) some training in communications skills so that project team members can implement partnering communications procedures, goals, and conflict resolution processes. The training need not be extensive; a little bit can go a long way in enhancing partnering effectiveness.

9. *Partnering involves a strategic mix of tasks, strategies, and skills.* Partnering blends a wide range of improvement efforts into a comprehensive whole. The Goals Statement involves a different kind of work and produces quite different results than the communications procedures.

Using different tools builds insurance into partnering. If one fails to produce results, two more remain. In addition, the different tools of partnering differ in an organized way. The goals statement, communications procedures, and conflict resolution process tackle communications in a comprehensive way such that no one of them can be accomplished alone.

10. *Partnering emphasizes follow through.* The two fields of design and organization development share a common strength and a common problem. Both are recognized for their quality of thinking, ideas, and creativity. Both are criticized for their inability to translate insight into consistent action.

Architects are criticized for isolating design from user needs and satisfaction and for separating design from the people who live and work in the completed building. Organization development programs are criticized because they are frequently dropped after an initial round of fanfare and superficial training. Since partnering applies organization development tools in the fields of architecture and construction, the issue of follow through looms as a concern for its effectiveness.

Partnering attempts to deal with follow through by building partnering workshops and tasks into building plans from the very outset, beginning with bid letters and requests for proposals. Since its origins, partnering has been designed to follow a project through to completion.

Even with this, however, follow through is a concern. Although partnering history and lore emphasize it, the inevitable emergencies of project life make it difficult to follow through on follow-through activities. At least partnering has the intentions and history with follow-through to give it a fighting chance to bring good ideas into the everyday reality of project life.

See Fig. 5.2 for an overall view on why partnering works.

6
Problems

Summary: External and Internal Problems

"Our partnering didn't work," some people say. Others generalize, "Partnering doesn't work."

When partnering efforts fall short, the causes can lie in two quite different areas (see Fig. 6.1):

- Mistakes people made in implementing partnering
- Dilemmas and problems inherent in the idea of partnering

Most of the time, the problem is not with partnering itself at all but with the way people implement it. Specifically, people try to shorten the process and leave a step out. To remedy these kinds of problems in the execution of partnering, we point out the most frequent mistakes we see people making in implementing partnering.

The dilemmas built into partnering are more subtle and require more subtle handling. For these, we illustrate specific dilemmas and suggest strategies for minimizing their impact.

Whether the problems are caused by the ways people implement partnering or dilemmas inherent in the very nature of partnering, addressing them is a major task for improving partnering effectiveness. Partnering will become more effective as participants build an awareness of its problems and take intelligent action to address them.

Problems *external to partnering*

- Implementing and following through on

  Original partnering documents
  Preparation interviews
  Documents (goals, communications, and conflict proce-
  dures)
  New practices and meetings
  New everyday behaviors

- Overemphasizing the initial workshop
- Providing inadequate communications skills training
- Not having the right people participate

Dilemmas *inherent in partnering*

- The *labeling dilemma* (We can label it "partnering," but we cannot control how others use the label.)
- The *selfish partner dilemma* ("If you were really my partner, you'd let me do what I want.")
- The *conflict avoidance dilemma* ("If I were really your partner, I would let you do what you want.")
- The *canned change dilemma* (How to keep programmed change fresh.)
- The *open discussion dilemma* (Let sleeping dogs lie?)
- The *theater dilemma* (Using partnering as a stage to publicize individually focused concerns)

Figure 6.1. Problems that impact partnering.

External Problems: Understandable Implementation Mistakes

Partnering is a straightforward process with fairly simple guidelines:

1. Prepare for partnering by building it into the contract. Prepare for partnering workshops by interviewing group members before the session.
2. Conduct a preconstruction workshop at which the group writes, signs, and plans to implement the following:
   - *Goals statement*
   - *Communications procedures*
   - *Conflict resolution process*
3. Provide training and facilitation so project team members get to know each other as individuals, build trust, and learn to resolve their own conflicts.
4. Conduct follow-up workshops at appropriate intervals to monitor partnering effectiveness, address current issues, and resolve problems.

People make mistakes by not following these guidelines. They neglect one task, overemphasize another, and perhaps attempt to skip one completely. These mistakes are understandable because often the people who make them don't have enough experience with partnering to know what the impact of their decisions will be.

Inadequate Follow Through

This is by far the most frequent mistake people make in implementing partnering. It usually occurs at transition points. People often fall short on bringing an idea or commitment from one phase of partnering to the next.

Preparation Documents. Preconstruction bid letters attempt to explain to bidders what will be expected of them to participate in partnering workshops. When the actual workshop occurs, however, it may not fit the original description. People in the workshop complain, "This isn't what the documents describing partnering said we would be doing."

This is an understandable problem because a considerable amount of time may elapse between the writing of prebid letters and the actual workshop. In fact, the people who wrote the original partnering outlines may well have left the organization before the partnering workshop occurs.

To remedy this problem, people who manage the partnering process must pay special attention to work done before they became involved.

Preparation Interviews.   While the facilitator may interview project team members before the workshop, the actual workshop content may not reflect well what every person brought out in the interview. People complain, "Why did you ask me all those questions and then not address them in the workshop?"

This also is an understandable problem. The facilitator builds the agenda based on his or her own judgments about key issues. Some individuals' concerns may get left out. Also, during the time the facilitator is conducting interviews, conditions on the project may change. Also, the facilitator may not make the best judgments about which issues to cover, or the facilitator may be swayed by key project team members to ignore an issue.

As this problem is primarily the facilitator's, it is up to the facilitator to address it. Skilled facilitators check the content of their agenda with group members in a tentative way before making the agenda. Many facilitators keep the agenda flexible at least in part over the course of an entire workshop day in order to respond to legitimate participant concerns.

Implementing Documents.   People may write outstanding goals statements, communications procedures, and conflict resolution processes but may fail to implement them adequately on the job site. This is a serious but understandable problem. Once people have finished writing a document they feel as if their work is done, especially if they have also signed it. Discussing the specifics of implementing seems redundant or abstract. The document seems to speak for itself.

As the facilitator is not present on the job site, it is the workshop participants who must deal with this problem. They are the ones who must live out their own promises.

To remedy this problem, it is helpful to devote sufficient time in the workshop to planning in a very specific way how the documents will be implemented. We follow the rule of thumb of allowing at least as much time to discuss implementing a document as to produce it. We try to take more time if possible.

Then, in discussing implementing the document, it is helpful to describe in a very specific way who will do what when. We actually often begin the discussion by putting those headings at the top of a flip chart page and steering the discussion towards filling in the boxes (see the sample in Table 6.1).

Implementing New Practices. Participants in the partnering workshop plan to implement new meetings, mechanisms to speed up information flow, or informal get-togethers but often fail to execute these plans. Other things come up, the participants forget to schedule, and the idea of the plan loses its sense of urgency.

As with failing to implement the partnering documents, the project team members must deal with this problem also. Unlike with the project documents, though, this problem can be elusive. Without something tangible like a document, it can be difficult to pin down exactly what the meeting or mechanism was supposed to be. If the partnering workshop succeeds in lowering group members' general tension level, it may also diminish their feeling that it is necessary to act on some of the more intangible promises they made.

Again, to remedy this problem it is helpful to get very specific in the workshop about who will do what when. A checklist such as that shown in Table 6.1 would help.

Changing Day-to-Day Behavior. A thoughtless, abrasive, off-the-cuff comment can undo, in a few moments, many hours of a group's work in partnering.

For a week or so after the partnering workshop people treat each other well, make an effort to listen to one another, and resolve conflicts cooperatively. Then something happens that sets someone off and individual behavior reverts back to prepartner-

Table 6.1   Implementing a Goals Statement

Sample Planning Grid to Enhance Implementing Partnering
Documents

Who	What	When
Architect	Makes copies of original flip chart for all project team members plus ten additional copies; sends to them	Within one week from now
Architect	Discusses goals at conclusion of regular weekly project meeting	Within ten days from now—next project meeting
Site supervisor	Posts copy in trailer; discusses with subordinates on site	Within two weeks from now
Owner	Brings copy back to the office staff; discusses implications for the staff members and how they should communicate with project staff	Within two weeks from now
Electrical engineer	Brings copy to staff on project; discusses implications for staff on this project	Within two weeks from now
General contractor, Architect, Owner team	Explores ways to keep goals in front of team: mugs, T-shirts, memo blanks, etc.	Make recommendations to whole project team in two weeks

ing days. Tempers flare, conflict escalates, and it may seem as if partnering never occurred.

Of all the areas in which follow through is difficult, changing individual day-to-day behavior is the most difficult and probably the most important for the overall success of partnering. If members of a project team write outstanding goals statements, communications procedures, and conflict resolution documents but treat one another with disrespect, partnering will fail.

It is easy to understand why individual behavioral change is the most difficult problem of follow through. It is possible for people to write a document, sign it, attend a meeting, and follow a set of prescribed steps to hold an organized discussion because these spell out the details of changed behavior in very specific terms. In addition, implementing a document requires only a finite, limited time period in which people have to monitor their own behavior. Everyday project life is much more encompassing and so is much more likely to encounter problems.

To remedy this problem it is helpful as in the instances above to anticipate specific areas in which people are likely to encounter problems and plan ahead as to who will handle them. The formal conflict resolution process should help people anticipate areas in which they are likely to encounter friction and provide them with some skills to resolve them when they occur. It is in this area that the importance of communications skills training becomes most evident and urgent.

Postponing or Canceling Follow-Up Partnering

The initial partnering workshop is often an emotional event. People discuss past problems, vent past frustrations, set lofty goals, and learn ideal communications skills, all the while discussing a building that has not yet been started.

It is easy to overemphasize the initial workshop. Often it is the first program participants have attended. Often, it feels to participants as if they have accomplished a great deal and solved whatever problems are likely to occur such that it seems that no problems could possibly occur after all this work.

People consistently fail to maintain partnering. They've devoted considerable effort to an initial workshop and produced thoughtful goals and procedures. They've learned skills. As the project goes on, they may not see any pressing communications problem. Everything seems to be going along all right.

We hear, "Do we really need this follow-up workshop? I could really use the time to get some work done on the project. Can't we put the program off until we have something to work on?"

It is of course difficult to schedule time to work on something when no problem is apparent, especially when there will always be some last-minute emergency that *must* be tended to on the project. Handling communications in this way, however, is like maintaining a car only when it breaks down. The scheduled, preventive approach is much less expensive, easier to manage, and more conducive to more creativity.

On the practical side, one of the most effective ways to avoid postponing follow-up workshops is to make sure there is a fair cancellation policy for the facilitator. The standard practice in the field of training and development is to pay facilitators a full fee for programs canceled or postponed within a week or less of notice and a half fee for programs canceled or postponed with one to two weeks notice. This kind of clause treats the facilitator fairly and makes project team members think twice before postponing unnecessarily.

Why People Fail to Follow Through

Because failing to follow through is a recurring problem, we thought it would be useful to explore further why it occurs. A deeper understanding of the reasons for lack of follow through should help formulate action ideas to remedy the problem. See Table 6.2 for a list of these ideas.

Initial Workshop Success. When people work hard and feel they have achieved success in one phase of partnering work, it is easy to underestimate the gravity of the next task. Initial success makes it seem that there really isn't much left to do, and if there is, it will go at least as smoothly as the current one.

Initial Success on the Job Site. If the job goes well for the first few weeks after a partnering workshop, it is difficult to see that any problems exist and therefore easy to be lulled into a frame of mind that the initial workshop solved everything.

Reactive Culture of the Trades and Professions. The design and construction fields are used to management by crisis. Workers in these fields are very good at responding to crises but not so good at working proactively to avert a crisis.

Table 6.2 Resolving Partnering Follow-Through Problems

Problem	Reasons	Possible action steps
Translating prebid documents into actual partnering agendas.	Initial workshop success (follow-up meeting seems unnecessary).	Schedule follow-up parameters in the prebid discussions.
Translating prework-shop interviews into workshop agendas.	Initial success of partnering on the job site.	Devote some time in every workshop to planning the next steps in a very specific way.
Implementing partnering documents (goals statement, communications procedures, and conflict resolution).	Crisis orientation of the trades and professions. Conflict avoidance. Abstraction (planning for implementation is less tangible than creating a document).	Allot adequate time for planning follow-up meetings in workshops. Allow as much time to discuss implementing an idea as it takes to write it or agree on it.
Implementing partnering procedures and meetings.	Real change is difficult for organizations and for individuals.	
Changing everyday behavior and habits.		Plan follow-up activities specifically as given in Table 6.1, noting who will do what and when.
Canceling or postponing follow-up workshops.		Use a fair facilitator cancellation policy to discourage thoughtless postponing of follow-through workshops.

If there is no crisis, it may feel to project team members as if there is no real need to have a partnering follow-up meeting. On the other hand, if there is a crisis, then it is necessary to postpone the follow-up workshop in order to attend to the crisis.

When it is time to schedule partnering follow-up meetings, crises of "real" work have often evolved, making it understandably difficult to take time out for a workshop. Of course, if people don't take time for a workshop they will probably resolve the immediate crisis but fail to devise system improvements to prevent similar new crises from recurring.

Abstraction.   At the point of completing one partnering task, where the current success is tangible, the next task is abstract. Planning to implement a goals statement involves working with things that may or may not happen, while the goals statement itself is a document people can hold in their hands.

Technical Skill Backgrounds.   If quality circle facilitators were asked to estimate building costs, they would probably not achieve high rates of accuracy. People who work in design and construction firms typically have little formal training and experience in facilitating groups, implementing change, or working with people through difficult transitions. Thus their estimates for the amount and content of work involved in these tasks are not likely to be very accurate.

More specifically, we notice that people with technical backgrounds tend to underestimate the time and effort necessary for group and change management tasks. They understand and anticipate nuances of details that must accompany a technical task but do not see what must be done to work with people.

Conflict Avoidance.   This is an ironic but real problem: People who seem by their everyday interactions to thrive on conflict can shrink back from discussing real differences of opinion. Frequently when conflicts come up on a job, people are embarrassed by them, or they may become unrealistic and hope that "things will work out by themselves."

When we interview project team members to prepare for a follow-up partnering workshop, they often begin by reassuring us that there are no problems, that we did such a great job in the initial workshop, etc. (The facilitator in the Benson Management Services Case described in Chap. 14 later encounters this phenomenon.) Only after some discussion do people finally volunteer that perhaps they have "just a little problem, nothing major" to discuss in a follow-up workshop.

Real Change Is Difficult.   Follow through of any kind, whether it involves partnering documents or day-to-day behaviors, involves real change and real change is difficult. The literature on organizational change (as noted in Chap. 13, "Electronic

Partnering," in Part 3) is for the most part a literature about failures. When organizations attempt to change, many fail or fall short of complete success.

On an individual level the track record is not much better. If even a small percentage of the people who resolved to lose weight, exercise, stop smoking, or quit procrastinating were able to translate their intentions into real changes in behavior, a major national movement would occur. Any kind of real change, even change that people can see is useful, helpful, or beneficial, is still difficult to translate into everyday actions.

Inadequate Communications Skills Training

After follow-through difficulties, the next most common problem we encounter with partnering efforts is that they fail to include adequate communications skills training. Project team members need communications skills for two reasons:

- To implement and follow through on their formal partnering agreements
- To communicate day to day in a style and with sufficient skills to resolve their own conflicts

No matter how clear and wise a project's goals statement is, people are likely to disagree about the most effective ways to implement it. At that point, they need a set of communications skills that their professional and trades training did not likely provide.

Along with formal goals statements and other procedures, partnering implies that project team members will be able to communicate with one another in a reasonable way and to resolve whatever conflicts they have in a collaborative manner. These are sizable (and somewhat questionable) assumptions for people who, for the most part, have had little or no communications skills training and are working in the very business that contained enough conflicts to incubate the idea of partnering in the first place.

In Chaps. 10 and 11 we describe the types of training that are most important to support a partnering effort. In this chapter, in the section titled Problems, we want to underscore the fact that too many partnering programs do not include sufficient time and effort to train participants in the basic skills necessary to carry out formal partnering agreements in their everyday communications.

On the other side of the spectrum, partnering workshop participants usually rate the communications skills training we provide as the most important aspect of the partnering work. They cite examples to let us know that they use it on the job. It's just before the partnering workshop that they occasionally question the value of training.

Communications skills training is easy to leave out. People use numerous reasons that sound logical to themselves:

- *"I had a course on this once already."* Communications skills are action skills like skiing or golf. Would these people say that one skiing or golf lesson taught them all they needed?

- *"I've gotten along without this so far, why do I need it now?"* People probably don't need strong collaborative communications skills to survive in design and construction; the old win–lose, "I'm OK; you're not" may still carry people a bit. However, achieving success in any aspect of the field that requires extensive work with people does demand more sophisticated skills than those one can acquire on a job site.

- *"Nobody else does this."* Communications skills training is a major part of the professional and management development training curriculum at most corporations, including major design and construction clients. Because of their project-based finances and their ongoing financial difficulties, design and construction firms lag behind other kinds of businesses in providing high-quality training and development for employees. Thus people in firms may have an unrealistic view of how much training managers and professionals in other fields get.

- *"You can send me, but I'll never change."* Communications skills training does not attempt to change anyone's personality, only to enable people to achieve the goals they would like to with a communications style that supports those goals.

Who Attends

Attendance at a partnering workshop can create problems for the effectiveness of partnering in two ways, and it is easy to err in either direction:

- If the list does not go far enough into the decision-making hierarchy of the organizations that constitute the project team, the workshop group may make decisions that those senior people later actively veto.

- If the list does not go deep enough into the working group of people on the job site, the workshop group may make decisions that those on-site people later quietly ignore.

It is easy to fail at getting organizations' senior managers and decision makers to participate:

- They are busy.
- They say they understand and support partnering and don't need to attend.
- They say they want to "empower" their staff.
- They say they really don't have much to do with the project at any everyday kind of level anyway.
- The people who are charged with getting decision makers to participate in partnering may be the decision makers' subordinates and may have a difficult time being assertive with their bosses.

Yet if senior management decision makers do not actively participate in partnering, it is easy for any one of them to undo a whole workshop's worth of partnering effort with a casual nod.

In practice, we encounter the greatest difficulty in this area not with members of the professions or trades. Their senior partners or owners are usually willing to participate actively in partnering. The problem lies more frequently with client organizations, especially with large government agencies or corporations: the agency director, the judge, or the vice president of finance.

In these kinds of situations, the senior management client decision maker may not understand how design and construction works, may operate in a political manner, or really may be terribly busy. This person endorses partnering "in principle," perhaps even writes a note describing his or her support for partnering. People from the organization may bring the note to a partnering workshop, circulate it, even read it aloud and usually everyone at the workshop is happy.

The trouble does not occur until later. The senior person usually has the power to let the project get well into construction and then takes a casual look at the plans or walks the job site and issues a directive to the project team: "Change the design to give my secretary another window, and by the way, cancel those weekly project meetings [the ones the project team planned in partnering workshops] so you can get more real work done. And while you're at it, get that goals statement out of the trailer, it's confusing the drywall workers."

At the other end of the spectrum, it is also possible to turn partnering into a cozy senior management meeting with no substance. Architecture and engineering firm owners who have not actually done any design work for years because they market full time meet with client managers who are many organizational layers above the people who will actually occupy a building. They hammer out agreements, get to know one another, and then move on.

When senior managers' partnering is over, the group has usually accomplished one objective that may actually help the project later on. They have established strong enough personal bonds so that if a serious problem evolves and is escalated up the hierarchy later on, the senior people from the organizations involved can make effective telephone contact for a first round of attempts at resolving the conflict.

On the other hand, the "senior managers' club" approach to partnering usually has limited impact on the everyday tone and communications on the job. When senior managers finish their workshop, they hand the directives over to the people who actually have to work together on the job and who, understandably, may be less than thrilled about implementing them.

It is never possible to provide a definite answer to the question, "Who should participate in partnering?" because every job is different, but one useful guideline is:

> In some way, actively involve anyone who may possibly become a roadblock or obstacle later on.

This does not necessarily mean that everyone attends the workshop, but it does mean that people who can later block partnering results:

- Receive accurate, detailed information about how partnering will work.
- Get an opportunity to express their concerns about partnering.
- Provide information to the partnering effort about their concerns about the project. The wording here is important; it is not enough to provide them simply with an opportunity to express their views: that approach is too passive and could still lead to unwelcome surprises later on. Instead, it is important to make sure to elicit some input from them about their concerns.
- Should designate one of the workshop participants to communicate with them if they do not attend themselves. Working through a spokesperson makes it possible to provide people with accurate information as well as a "live" outlet to express their views.

At the senior management end of the spectrum, the suggestion to involve anyone actively who could later become a problem may mean providing a great deal of information to people who might otherwise be left out of partnering. People in government agencies worry over this rule of thumb. "If we involved everyone who might become an obstacle," they say, "we could have half of the agency here."

Of course it would be impossible to have all these people participate in a partnering workshop, but expressing such a concern outlines the degree of groundwork necessary to make partnering successful. If in fact a large number of people in the client organization could block or undercut the results of a partnering workshop, then it is all the more important to make sure that those people are informed of what partnering will do and are provided with an opportunity to participate early on.

A corollary to the rule of thumb noted above involves people who would be awkward or difficult to work with in partnering. If people express pleasure that certain potential partnering participants are not interested or unable to attend a workshop, it is usually a good sign that these are the same people to worry about undermining the work of partnering later on. We usually prefer to involve "difficult" people actively in partnering at the outset instead of proceeding without them with the chance of their causing more serious difficulties later on.

At the other end of the spectrum, to avoid the senior management club problem, we simply attempt to ensure that a vertical cross section of project team members participates in partnering. If the rule of thumb of including anyone who might undercut or block partnering is used, then it is usually necessary to include enough people who are active on the job site to bring partnering efforts to the project's everyday experience.

Dilemmas Inherent in the Partnering Process

While follow-through, training, and participation problems have more to do with the judgment and frailties of individuals, another kind of problem with partnering has more to do with partnering itself. Some problems are inherent in the nature of partnering. Follow-through activity and the other problems will vary depending on the wisdom, insight, and skill of the people involved in partnering, but the inherent dilemmas are always present.

Just as there are benefits inherent in bringing the project team together for partnering discussions and training, there are also inherent dilemmas. We describe these problems because the best remedy for them is awareness. The more partnering participants are aware of these dilemmas, the more they can minimize their impacts.

The Labeling Dilemma.   This is a general version of several more specific dilemmas. People use the label *partnering* to refer to all sorts of other activities and intentions that really don't have much at all to do with the kind of partnering described throughout this text.

Once a label is attached to a process of organizational change it is not possible to control how people use the label. Calling the effort *partnering* invokes a *halo effect* and a *Hawthorne effect*, causing improvements in project team members' attention to communications beyond the actual events and scope of a partnering workshop.

Calling the effort *partnering* also causes problems of rhetoric and mislabeling. Specifically, people can inaccurately or wrongly "invoke" the name of partnering when they are really doing something that conflicts with the original intentions of partnering. People can attribute negative behavior to partnering as well as positive ones.

The Selfish Dilemma. *"If you were really partnering, you'd let me do what I want."* This is one more specific way people misuse the label of partnering. In this situation, project team members accuse others who have a legitimate disagreement with them of not participating in the project in the "spirit of partnering." Of course partnering encourages healthy disagreement, but the partnering label can be a convenient weapon to use against someone who is arguing an inconvenient point.

The Conflict Avoidance Dilemma. *"If we were really partnering, I would let you do what you want."* Sometimes people who do not fully understand partnering use the label to avoid conflict. They think that partnering means avoiding conflict, that conflict is a bad thing.

Ironically, people learn skills to resolve conflicts, write procedures to manage them, and then fail to bring them to the surface when appropriate. So much time and effort are devoted to miscommunications and conflict that people may get the message that somehow they should not be disagreeing. Even if the content of a partnering workshop includes a discussion of the positive values of conflict and disagreement, the fact that everyone is taking so much time to work on conflict can make some people less willing to discuss it.

The Canned Change Dilemma. Partnering encounters the same dilemma as total quality management, just-in-time inventory control or any other "canned," prepackaged approach to organizational change. All these efforts are worthy, sensible approaches to organizational improvement, but the fact that they are "canned" poses a challenge to their continuing freshness, relevance, and usefulness.

The chief problem with canned change is that it provides an answer before the problem is fully defined. It provides solutions before people fully comprehend the issues.

Canned change programs have some advantages:

- They provide a neat, coherent packaging of what could be a sloppy process.
- They are more easily understood than more complex, grass-roots efforts.
- They can generate more specific, measurable results.
- They are more easily managed.

The problems of canned change programs include the following:

- The solutions they provide may address problems that are insignificant because those problems are part of the canned package.
- They may "miss" major problems while focusing on minor ones that happen to fit the canned package.
- Their neatness can lull participants into a sense of complacency.
- The action of diagnosing problems missing in canned programs builds in interest and commitment to following through on the results and solutions.

As a "canned" change effort, partnering benefits from the strengths and suffers from the problems listed above. In practice, partnering simplifies complex issues enough to make it possible to work on them and make real progress.

The Open Discussion Dilemma. This is more myth than dilemma, but it is a myth prevalent enough to impact the behavior of some people who participate in partnering: "If we talk about this problem openly, it will make the problem larger than it really is and potentially, more difficult to resolve." Most people who participate agree with the partnering stance: "In order for problems to be resolved, they must be discussed. Discussing a problem does not create or enlarge it, it simply makes it tangible so that it can be addressed."

Still, at partnering workshops, people don't raise serious issues until several hours into the session. People discuss some issues in the early hours, but hold back on something, often something of consequence, until near the end of the workshop.

No research documents the "If we discuss it, it will grow" myth, and in some ways, it has the ring more of an old folk tale than of a current, widely held belief. However, we usually encounter at least one person on every project who holds this belief and argues for it. We also encounter several people on each project who may not articulate this belief but act as if they support it.

The Theater Dilemma.   Partnering levels the playing field of discussion on a project by creating time and space so that all participants can truly participate. This may help the underdogs in the process to express their views more effectively, but it also provides a stage for people who may bring information that is irrelevant, selfish, or one-sided.

Principal members may participate in a partnering session to represent their firm even though they are removed from the day-to-day concerns of the project and perhaps even misinformed as to the real concerns most impacting successful job completion.

Accusing a project team member of errors and problems, people can also create the appearance that the team member is not performing up to par. When the accused person returns to his or her office, he or she may discover that the accusations were not based in fact.

Beyond these aspects of people intentionally misusing the forum partnering provides, there is also the more basic problem of crowd control. People who work in the building professions and trades have a well-earned reputation for being strong willed, highly opinionated, and inflexible to opposition. Without trying, these are people who can easily turn a simple difference of opinion into a battle royal. A roomful of such people presents a challenge even to the most experienced facilitator.

2
Partnering Documents

Three written documents define the partnering process, providing tangible products and benchmarks to guide follow-through activities. This part provides detailed guidance for writing and implementing the three documents. We stress implementing because writing the document is only half of the task, usually the easier half. We provide numerous specific examples and templates of documents based on real partnering projects.

7
The Goals Statement

Summary

Project team members work together to produce a *goals statement*, a listing of their individual goals for the project. In writing the statement, they usually discover that they share many more common concerns and agree on much more than they may have thought.

Writing the goals statement is just the beginning, however. The project team must also work hard to bring the statement to life in two ways: implementing it in such a way that it impacts everyday interactions on the project and using it as a yardstick to monitor project performance.

What the Goals Statement Is

The goals statement is one of the most visible outcomes of a partnering workshop. The statement is a listing of the goals project team members agree on to guide the project. Along with the communications procedures and issue resolution process, the goals statement is one of the three tangible partnering products.

As with the other tangible partnering products, the real value of the goals statement is not in the product itself but in the way it is used. Writing the goals statement is only one-third of the work

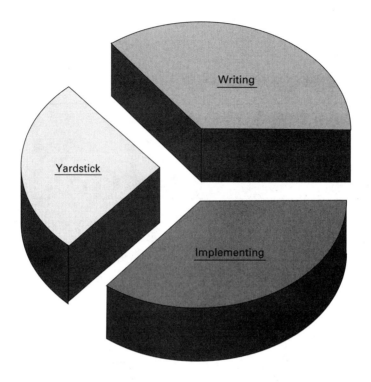

Writing the goals statement is only a third of the effort of working with it. The other two thirds come from
 • Implementing the statement on the job site
 • Using the statement as a yardstick to measure ongoing project performance

Figure 7.1. Working with the goals statement.

involved in using the statement. The remaining (and more difficult) two-thirds involves implementing the goals statement on the job site and using it as a continuing yardstick to monitor job and partnering performance. (See Fig. 7.1.)

To write the goals statement, each project team member lists a goal that is important to him or her and presents that goal to the group. If the group agrees, the goal is included in the list. The group may also suggest refinements and/or modifications. Groups may reject a goal. Most often, though, they work in a spirit of cooperation to include goals their members believe to be important.

Typically, a goals statement includes between 10 and 20 goals listed as *bullets*. (See Figs. 7.2 and 7.3.) Since the list results from individual concerns, it usually includes a wide range of concerns. Some are very practical: "Finish the job ahead of schedule." "Eliminate unnecessary costs." Some are more idealistic, focusing on communications and relationships: "Treat each other with respect and consideration at all times." "Always work to the best of your ability." Some goals may be very personal: "Have fun." "Enjoy the work."

Alternatively, it is possible to construct the goals statement from a series of requests project team members make of each other. For example, the contractor may ask the architect, "I would like to get back some response to all my questions…not necessarily a complete answer, but some recognition that you got the

---

- Stay on or ahead of schedule.
- Treat everyone on the project with respect.
- Always do the best job possible; "quality is free."
- If you have a complaint with somebody, talk to them directly. Don't go behind their back.
- Make sure you communicate the information others need to know.
- Don't spend money if you don't have to.
- Try to settle arguments before you escalate them.
- Handle conflicts by working on them, not avoiding them.
- Return all messages before you go home for the day.
- Turn information requests around in 24 hours or less.
- Make a profit: we all have to succeed.
- Put your energy into solutions that work for everybody, not just for you at someone else's expense.
- Look for opportunities to innovate and improve quality.
- Have fun! We chose this work in the first place, it's up to us to make it enjoyable.

Figure 7.2. Sample partnering goals statement (list form).

I, the Architect, will:

- Return all contractor's calls within 24 hours.
- Accurately communicate engineer's recommendations to the group in a timely and complete way.
- Provide the whole project team each week with a current running list of the status of all requests for information.
- Keep my own project team informed of changes in job status that affect them.
- Spend the time necessary to explain drawings to people working on them.
- Make myself available to all people working on the job who need information I have.

I, the contractor, will:

- Communicate my requests as requests rather than as demands.
- Listen long and hard before losing my temper.
- Make myself accessible on the site with beepers or whatever it takes.
- Not sacrifice job quality for the schedule.
- Not sacrifice the schedule for my own stubbornness.
- Work as hard to resolve conflicts as to stick up for my own opinion.

Figure 7.3. Sample goals statement (question–response form).

request…within a day. Do you think you can do that?" The architect's response, "I will always in some way get back to you within 24 hours," is listed as a goal.

If the person responding to a request doesn't think it will be possible to meet the request, he or she may suggest a slight modification to the request. For example, the architect in the instance above may hesitate to commit to 24 hours but agree to 48 hours as a deadline for providing responses. If the issue is very important, the two people may work for some time to devise a request and response that are acceptable to both.

In either case, once the goals statement is written, the project team devotes substantial time and effort to plan how to implement it. They discuss predictable roadblocks and how to overcome them. They plan how to communicate the statement to people on the job site. They discuss how to implement the statement so it will influence their own actions.

Finally, the project team plans how to use the goals statement as a yardstick. If the statement really does represent the team's hopes for the project, it can be used at later partnering workshops to gauge project and partnering performance, to identify problems, and even to suggest how to go about finding solutions.

What the Goals Statement Can Do

If written and implemented effectively, the goals statement produces different outcomes and results in three different areas of partnering: in the initial workshop in which the project team writes the statement, on the job site, and in later partnering workshops.

The Goals Statement in the Workshop

In the workshop in which the project team writes the goals statement, the statement provides focus, direction, an early success, and a motivational experience. People attending the workshop may have some misgivings about the project, each other, or even about the workshop itself. Yet they will seldom disagree on goals because they all usually want the same kinds of things: cooperation, accountability, clear communications, a reasonable effort, quality work, follow through, and responsibility.

Working together on the goals statement can help the diverse members of the project team see that their goals are not so diverse at all. Because it is usually easy to agree on goals, working on the statement provides an early success at partnering work. Workshop planners frequently schedule work on the goals statement sometime near the beginning of a workshop because it creates a positive tone and motivates the group for the more difficult work of establishing communications procedures.

The goals statement also helps a workshop group by providing a tangible outcome of partnering. Workshop participants are usually most familiar with tangible entities: concrete, conduit, working drawings, and cost figures, for example. One of the aspects of partnering that can be most unnerving to them is partnering's intangibility. The completed goals statement is written in bold marker pens and taped to a wall where all project team members can scan it, a tangible outcome of an intangible process. They frequently continue to scan, scrutinize, and reflect on it over the course of the rest of the workshop day.

Beyond the motivational and emotional impact of the goals statement in the workshop, the statement also provides important direction, focus, and structure for the workshop and for the other partnering products: communications procedures and issue resolution. If the project team writes a goals statement that is thoughtful, clear, and complete, it is then much easier to agree on communications procedures and issue resolution.

The Goals Statement on the Job Site

The goals statement is supposed to impact everyday behavior on the job site, yet many (too many) goals statements literally never leave the workshop room. If participants leave the goals statement taped to the wall of the workshop room when the partnering workshop is over, it is a sign that the goals are unlikely to influence the job.

It is fairly easy to write a nice, even a wise and thoughtful goals statement in the confines of a partnering workshop. The workshop is usually held in a quiet, comfortable hotel or office, often long before the real pressures and tensions of the job become evident. Implementing the goals statement in the real world is another story.

Bringing the goals statement back to the job need not be time consuming or difficult, it just takes a bit of thoughtful effort. Here is where the project team can get creative in its thinking. Project team members can bring the goals statement back to the job in many ways:

- Posting the actual statement in the construction trailer or some other public place on the job site
- Printing the statement on notepads, posters, and even T-shirts and coffee mugs used on the job
- Discussing the statement at regular project job meetings (even if this is brief it can be effective)
- Inserting the goals statement in pay envelopes

People on the job site are usually interested in what went on at "that workshop that all the foremen went to." The goals statement helps show people on the job what was discussed and how the members of the project team hope to communicate on the project. The statement also outlines expectations for how things will go on the job. It is not at all uncommon to hear craftspeople and subcontractors who did not attend the partnering workshop use a goals statement to manage communications on the job.

The Goals Statement as a Yardstick for Project Performance

The goals statement serves an important purpose in follow-up partnering workshops when it is unrolled, taped to the wall, and used to assess project performance. The facilitator can poll the project team and ask them to score the project's effectiveness thus far for each goal, assigning a value of 1 (low) to 5 (high). The facilitator then posts each team member's scoring for each goal on the flip chart. If a number of people give a goal a low score, then that becomes an issue for discussion and problem solving.

For example, if several members of the project team score the goal "Respond to all requests for information within 24 hours" on the low side, then the group discusses how requests for information are going. Typically, the facilitator asks participants to list positive and negative points, i.e., things that are going well or not well for that particular issue. In this case it may be initially that the contractor is complaining that the architect is slow to respond to requests. Upon further discussion, though, the architect may be complaining that the contractor is asking questions in a way that is ambiguous and unclear. In any case, the group works to define the problem and to outline suggestions for action steps to resolve it.

Using the goals statement as a yardstick in this way provides an important focus and grounding for partnering. In a follow-up workshop, using the goals statement provides a sense of continuity and enables participants to avoid starting from scratch every time there is a problem. The goals statement also provides a useful tool to resolve many issues. Often the solution for a particular problem is a goal that the group has already articulated. In the case above, for example, the group has already agreed to the important baseline of 24-hour turnaround.

Larger Potentials of the Goals Statement

Vision, mission, and goals statements hang conspicuously on the walls of most of the companies we work in these days and with good reason. If these documents are put to use, they can provide focus, guidance, and structure for an organization. They can set a tone for how people treat each other and work together. Or they can have as little impact as the corporate art with which they usually share wall space.

The goals statement of a partnering project resembles and shares the potential and problems of the vision, mission, and goals statements that are fashionable in all kinds of organizations. Like those statements, a partnering goals statement can provide focus and guidance and establish a tone for the project. Like those statements, a partnering goals statement can make direct gains in improving project performance. The key to the impact of the goals statement is usually not in the statement. Of course it doesn't hurt to have a well-written statement, but what's more important is how the statement is used.

Beyond its direct impact, the partnering goals statement also has the potential to achieve four very special outcomes: create a project culture that supports productive work, generate creative solutions to old problems, increase overall accountability, and improve organizational learning. These outcomes result from the roots of the statement in design and construction.

Create a Positive Project Culture.   Every construction project is an organization in itself, each with its own culture. On several different job sites of similar projects, one is likely to encounter vastly different cultures. Some are cooperative and supportive. Some are cold and impersonal. Some are combative, war zones of conflict and disagreement.

Every new project brings with it the opportunity to create a strong, positive project culture. It is not an easy task; there are so many variables and factors outside anyone's control: costs, climate, local conditions, personal histories, the divided nature of the design and construction business.

At the least, writing a goals statement helps remind project team members that they do have some level of influence in making the job what they want. The statement provides one tool for creating the project culture.

Creative Solutions to Old Problems.   Working on the goals statement, the facilitator can encourage project team members to keep raising their expectations, aiming for more creative solutions to old problems. One classic case of this creativity involves a health care residence.

Initially, the facility manager just asked the project team to try to be decent to the facility's current residents. As the team discussed it, however, they enlarged the goal. Why not treat the patients as honored guests? Why not do more thorough, serious outreach to involve them in the project? It would probably make the project go more quickly in the end, and the effort would make for great community relations.

Viewing the patients as resources rather than obstacles led to creative resolutions of numerous real job problems. The job fence, for example, usually an inconvenience at best or an eyesore at worst, became the canvas for a patients' art competition and mural.

Increase Overall Accountability.   On the face of it, it may seem that goals statements and accountability are at opposite ends of a spectrum. Accountability may seem more like policing or monitoring, and goals statements may seem more like wishing and hoping. Yet writing the goals statement serves the function of making project team members more accountable.

When they articulate their goals and write them down, project team members make a public statement about what they want for the project. Because the statement is made in public, it is always available to them and others as a reminder of what they set out to do.

The accountability outcomes of the goals statement are evident in a service station that posts its corporate values on small plaques on its the gas pumps: quick purchases, friendly service, clean restrooms, good neighbor, etc. By placing these goals in public view, the service station increases its accountability to them. Customers who may feel that the service station is not measuring up to one of the goals find it easier to communicate with the owner. The owner himself or herself is reminded on a daily basis of the kind of station he or she is trying to run.

The goals statement can work in a similar way on a construction job site if it is made available in a similar way. Making the statement public helps increase its value as a tool to increase accountability.

Increase Organizational Learning.   One of the most disturbing and frustrating aspects of design and construction work is not that people make mistakes but that they make the *same* mistakes, repeating them from job to job. Anyone who has worked with design and construction for just a few years has seen some ways of getting the job done that work better than others. Yet on most jobs there is no opportunity for people to raise the issue, discuss, and implement better practices they have seen and used on previous jobs.

Taking the time to write and implement a goals statement gives project team members the opportunity to stop and tap into their often substantial experience and use that experience in the new job. That kind of building on prior experience creates and represents organizational learning in the project. This kind of event may also represent learning for the different firms involved in a project as they have the opportunity to work with the ideas and experience of one another and use them for their own applications.

Improving Goals
Statement Effectiveness

Like any attempt to improve performance, the goals statement is subject to pitfalls and problems that may detract from its effec-

1. Get full participation:
   - In discussing the statement
   - In implementing the statement
2. Aim high; avoid a "business as usual" approach.
3. Sign the statement.
4. Take the time necessary to discuss implementation plans in depth.
5. View partnering as something that happens on the job, not in a workshop.
6. Build the goals statement into everyday project work; don't keep it on a wall:
   - *a.* Visually
   - *b.* In performance discussions
   - *c.* As an integral part of everyday job operations and procedures
7. Devote adequate effort to follow-up workshops.
8. Be specific in using the goals statement as a yardstick.

Figure 7.4.   Improving goals statement effectiveness.

tiveness. These differ for each of the three aspects of using the statement. It is possible to improve the effectiveness of each (see Fig. 7.4).

Improving the Effectiveness of Writing the Statement

1. *Make sure everyone participates in the process in an active way.* Since the goals statement is often written at the beginning of a partnering workshop, it may be that the people who are initially most comfortable participating in meetings dominate the writing of the document. This is understandable:

- The facilitator may hesitate to push to involve the more quiet participants, at least at the outset.
- The whole group may be feeling both enthusiasm and pressure to accomplish something tangible.

- People who are slow to participate may worry about slowing down the momentum of the group.
- People who are comfortable participating may be energized by the activity and find themselves producing numerous statements the group accepts.

Less than full and complete participation at this point is unfortunate, however. A goals statement that has maximum impact on the job site and in later partnering workshops will trace its roots to the active input of *all,* not just some or even most of the project team.

Full participation has two components: writing the document and fully discussing and agreeing to the goals others write. When people actively work at both, they are more committed to supporting the document later.

Moreover, people who may be slow to participate usually do have their own ideas, and those ideas are often substantially different, often deeper than those of a vocal minority. It would be a pity to leave them out.

Actively contributing to the writing of the document means that every project team member contributes at least one goal. If people are slow to define a goal, it is usually worthwhile to slow down, take a break, and perhaps ask each participant to write a thought down before proceeding. To take the pressure off presenting the idea further, people can discuss their goals in pairs or perhaps tape them to a chart at the front of the room.

Fully discussing the goals means that the whole group gets to review in substance and detail each goal that individuals present. The effort here is not to pick the goals apart but to settle on a goal that will be fully supported by the people who did not write it. Ways to increase the discussion, buy-in, and support of the goals suggested by others include:

- Checking with each member of the group for modifications and refinements while the goal is being posted
- Assigning each goal to a subgroup of two or three people to discuss its importance, anticipate the likely obstacles that will get in its way, and map out some strategies to overcome those obstacles
- Asking the whole group to select the goal they think is most important, most frustrating, most enjoyable, most different, most connected to their own satisfaction, etc.

The exact form the group takes to discuss the goals does not matter so much as the fact that they spend time and effort actively working and thinking about the goals. This is the kind of work necessary to make the goals statement a living document.

2. *Aim high; avoid a "business as usual" statement.* The same pressures that may skew participation in the goals discussion can also drive the group to come up with a goals statement that is complete but uninspiring. This may also occur if people are more used to working with goals as business projections than as aspects of a vision or mission.

The goals statement is not the place to reiterate simply the business requirements of the project: "on time, under budget, no rework," etc. The statement is an opportunity to aim a little higher and to make this project a little better than the ones that went before. If this project is to be different, the goals statement offers a tangible opportunity as a start.

3. *Sign the statement.* Having project team members sign the goals statement when they complete it brings a solid sense of closure to the activity. In addition, signing it reinforces the fact that the statement is intended to guide future interactions and work on the job site.

Improving the Effectiveness of Implementing the Goals Statement

1. *Take the time necessary to discuss implementation plans in depth.* Once the goals statement is written, it is tempting to consider it complete and move on to other partnering work. It is difficult for many people to discuss implementing it because that kind of discussion involves working with intangibles and possibilities. Many people who do design and construction work are more at home in the tangible world.

   Nonetheless it is important to think through how to implement the statement once it is written. Discussing implementation will surface new ideas as well as potential obstacles and snags. People may borrow some of the better ideas they hear from others and anticipate how to work through obstacles if they have a chance to talk the plan through.

2. *View partnering as something that happens on the job, not in a workshop.* One of the obstacles to implementing the goals statement is the frame of mind many project team members have that partnering is something that happens primarily in a workshop. To them, the goals statement is a good thing, but it really applies to the workshop, not to everyday project communications.

   Unfortunately there is often something about the way partnering is done that reinforces this gap. The session may take place in a location far from the job site, in a hotel conference room that differs significantly from any meeting room where the project team members will ever meet.

3. *Build the goals statement into everyday project work; don't keep it on a wall.* Every major corporation has goals, but few corporations use their goals effectively to drive the business or implement them in the everyday work life of employees. Companies typically get their goals off the walls of executive conference rooms and into the worklife of employees in three ways that design and construction projects can emulate:

   a. *Visually.* Companies find many ways to visually keep their goals in front of employees. Employees may find versions of company goals appearing on all meeting room walls, on key chains, on business cards, on computer screen savers, and in pay envelopes. While there may not be many finished walls available on a construction site, there are many places where the project goals can be printed and displayed: on coffee mugs, on T-shirts, in pay envelopes, on fences, on notepads, over coffee pots in the trailer, etc.

   b. *In performance discussions.* Many companies incorporate corporate goals as criteria on employees' performance reviews and appraisals. In some, salary and raises are directly linked to performance against goals. This is an effective way to implement goals but not easily transferable to design and construction, as performance appraisals and reviews are not always part of the way firms operate. Also, if they do have performance management programs, the different firms working on a job are likely to differ in the way they implement those programs.

   Nonetheless it is worthwhile discussing in the partnering workshop how the various firms involved might incorporate the project goals statement into individual performance feedback. Often workshop participants are people who can speak on behalf of the policies and procedures of their firms, and they are often interested in giving the goals statement an opportunity to influence job performance.

c. *As an integral part of everyday job operations and procedures.* This is the area where implementation of the goals statement can become very worthwhile and very creative, yet which is often overlooked and neglected.

Improving the Effectiveness of Using the Goals Statement as a Yardstick

1. *Devote adequate effort to follow-up workshops.* One of the major reasons people fail to use the goals statement optimally as a yardstick is in the larger problem that too many partnering efforts end after one workshop. In the same way that managers in every business struggle to meet their obligations in providing employees' performance appraisals, another measurement effort, partnering efforts suffer from gaps and shortfalls in follow-through activities.

Ironically, one of the reasons people fail to follow up in partnering lies in the initial success and energy generated in writing a goals statement. After producing a goals statement, discovering how much they agree, and generating a high level of energy, people on the project team feel as if they are finished. Riding a wave of consensus, agreement, and high energy, they have a hard time imagining that they will ever disagree, much less need to get together for a day of follow-through workshops.

2. *Be specific in using the goals statement as a yardstick.* Beyond failing to get together at all to follow through, the other problem people encounter in using the goals statement as a yardstick is a lack of specificity in working with it. For example, the facilitator in a follow-up workshop may post the goals statement on the wall and ask the group, "How are we doing in progress against these goals?"

This kind of general question will generate a great deal of discussion but the discussion will be problematic in several ways:

- Some (potentially unimportant) goals will get too much attention.

- Some (potentially important) goals will get neglected.
- Individuals' level of participation in the discussion may not indicate the strength of their sentiments, i.e., outspoken people may sound more concerned than they are, and people uncomfortable with participating may not effectively represent their own opinions.

A more specific and more useful discussion begins with having each member of the workshop score the effectiveness of the project on a five-point scale for each goal. The facilitator then records every person's score on a flip chart page. The group then scans the page and can easily identify the specific goals that are problem areas for the project.

Once the scores are posted it is usually useful to have the whole group discuss the scoring. However, in order to address the difficulties underlying the problematic goals, it is usually necessary to break the group down into smaller subgroups. Smaller groups are usually more effective in focusing on a problem. They develop action alternatives and present those alternatives back to the larger group for a final decision on a course of action.

First Person: Participant Comments on the Goals Statement

> I have to admit I was skeptical about this. I knew what the goals statement was; I had heard about them on other jobs. I just didn't see how it could have any real impact on a project, it all seemed so "touchy-feely."
>
> Now that we're into our project, though, I can see how the thing really works. It does make a difference in the way the project is going. It particularly makes a difference in the way people treat each other.
>
> The key to the whole thing, though, isn't what we wrote. The key to it all is that it's posted in the trailer, over the coffee pot. Everybody sees it, and we make a point of talking about it at project meetings.
>
> *On-Site Manager, General Contractor*

> I couldn't believe the energy that came out when we wrote the statement. I worked with some of these characters before, and I never saw anything like it.

I also couldn't believe how much we were all in agreement, this group of people who usually are fighting with each other. It was a very good thing to do.

Director of Real Estate, State Agency

Signing it was the thing that got to me most.

I thought it was all some "blue-sky" thing, but when the facilitator stood back from the flip chart, uncapped the marker, and went to give it to the project architect to start signing it, the whole feeling in the room changed. We could tell this was going to be serious.

Project Manager, Civil Engineering Firm

This whole project was different than anything I've ever worked on, and I've worked on a lot of them in the past 18 years. We didn't meet the schedule, we beat it, and we did it together. I never worked on a project with as few fights.

I think we set the tone for the whole thing in the goals statement. The facilitator pushed us to go beyond the everyday stuff. We had agreed on "meet schedule" as a goal, and he just shook his head and said, "With all the years of experience in this room, is that the best you can do?"

That kind of got to me, I guess, so I took up the challenge. I knew we could beat the schedule all along, so did half a dozen other guys in the room. Nobody wanted to say anything, though, because we all knew we would really have to work together in order to get it to happen. I took the first step and the others followed right away.

We did have to work harder on communicating in order to make it happen, but the results, a three-week savings on a one-year project, were more than worth it for all of us.

Project Architect, Architectural and Engineering Firm

Recommended Reading

Books on goals and goal setting crowd the shelves in the business sections of bookstores. One of them is especially good at conveying how a goals statement can influence the real actions of organizations and individuals. In *The Seven Habits of Highly Effective People* (Crown, New York, 1992), Stephen Covey writes with great clarity and purpose about how people can use personal goals statements to clarify what they want and drive real changes in behavior.

8
Communications Procedures

Summary

People frequently blame each other when miscommunications or missed communications occur on a project. They tell each other that they need to remember to circulate more information more often and in a more timely manner. They blame each other for keeping information to themselves. Sometimes they even accuse others of withholding information intentionally.

Although there may be some truth to all these complaints, whenever we hear them we usually discover that the major source of communications problems is not people but procedures. The problem is usually that people have not spelled out communications procedures in sufficient detail or worked with the procedures enough to make them effective.

Communications procedures help make communications more tangible. They are specific processes and rules people on the project team agree to follow in order to attain the goals they have identified in the goals statement. (See Fig. 8.1.) Participants in a partnering workshop typically devise procedures for two kinds of communications: meetings and written communications. Meetings include weekly, daily, and informal, as-needed meetings. Written communications include formal and informal procedures for clarifying issues as they arise on the job, clarification of authority and sign-off responsibility, and informal means of access in emergency situations.

1. Request for information (RFI) procedures
   - Formal
   - Informal
   - Documentation or log
2. Back-up access plans
   - Plan management
3. Project chain of command
   - Sign-off authority for making decisions
4. Meeting plans and procedures
   - Weekly project meetings
   - Daily "check-in" meetings
   - Decision-maker meetings
   - Informal meetings

Figure 8.1. Types of communications procedures.

Communications Procedures Follow from the Goals Statement

It is easy to get idealistic and hopeful about project communications in writing a *goals statement,* but more difficult to try actually to manage communications in detail with specific procedures. Devising communications procedures creates rules and processes for communications, intangibles that are widely misunderstood yet hotly debated.

Communications procedures make the perfect foil, the ideal counterpoint to the goals statement. While the goals statement gets project team members to push their heads into the clouds, communications procedures drag their feet back to solid ground.

The goals statement focuses on aiming high, thinking about possibilities, and planning for excellence; communications procedures calibrate the accuracy of the aiming process, focus on realities, and plan for consistency and detail.

Communications procedures provide the infrastructure through which goals are realized and attained. On the other hand, the goals statement paves the way for effective procedures by providing context and purpose. Without the goals statement, partnering communications procedures would be little more than a set of mindless rules, rules that project team members could break with little notice taken. Thus the goals statement also provides a means to help ensure that the procedures are followed.

How the Project Team Works on Devising Communications Procedures

In a partnering workshop, the shift from working on the goals statement to working on the communications procedures also marks a transition in mood. People move from the high energy and motivation that usually accompany writing the goals statement to a more serious, "let's get down to business" attitude. By the conclusion of the goals statement, some members of the project team show signs of discomfort with the emphasis on goals; they want to know how the goals are going to be converted into realities. Working on the communications procedures responds to their concerns.

The project team works on communications procedures in much the same way as they did on the goals statement. The facilitator usually begins with an open structure, asking each person to suggest one or two very specific procedures that would help the project attain the goals set out in the goals statement.

Project team members usually get their ideas from two sources: their own experience on successful projects and their own imagination. Thus writing communications procedures enables project team members to tap into their own experience and insight and to use them as a learning resource for current work.

Many people find these discussions are more productive if participants are asked to write their suggestions down on a piece of paper. Writing down the suggestions helps equalize participation, making it easier for quieter participants to remember their ideas and bring them into discussion. Writing down the ideas also helps preserve some of the innovative possibilities that can get lost in an energetic group discussion.)

Once everyone has written down a suggestion, the facilitator breaks the group into subgroups. (Again, breaking a large group into subgroups helps equalize participation and preserve fragile ideas.) The facilitator may ask each group of three or four people to aim to propose five or six suggested procedures to the whole group. Subgroups have 15 to 20 minutes to work on listing their ideas before presenting them to the whole group.

The product of this work on communications procedures is a list of very specific processes and activities that everyone on the project team agrees to follow. As with the goals statement, participants sign the final document in the workshop. Signatures are important to ensure that people take the procedures seriously and intend to follow them.

Because of the level of commitment and follow-through activities implied by the signatures, people on the project team tend to haggle over the procedures the subgroups present for approval. The discussion can easily become heated, occasionally to a point that people in the group wonder if this can be the same harmonious team that produced the goals statement.

For example, one person on the project team strongly favors a daily on-site meeting among all the key decision makers; another member of the team strongly believes that such a meeting would be a complete waste of time. One member of the team wants to have the architecture firm document all requests for information (RFIs) on a running log and have the firm fax the log to all key team members weekly. The architect on the team views the suggestion as an insult to her intelligence and integrity.

Because the haggling over communications procedures can become heated, it may threaten the viability of the workshop. Working on the goals statement enables project team members to identify areas of congruence and agreement, but working on communications procedures is more likely to raise areas of disagreement.

Haggling over communications procedures is inevitable, and it poses a challenge to a partnering workshop because it gives participants their first real problem to solve: how to forge an agreement in areas where they may strongly disagree.

Why Communications Procedures Involve Disagreements

Resolving the disagreements about communications procedures begins with a thorough understanding of why the disagreements occur in the first place. One reason why the debate over procedures threatens partnering is that people are unprepared for it. They seem to think that working on procedures is going to be very procedural: dry, analytical, noncontroversial, and almost boring. Understanding why discussing procedures generates conflicts helps resolve the disagreements that arise.

1. *The specificity of communications procedures offers more to argue about than the generality of goals statements.* Goals statements tend to result in general proclamations that offer little to disagree about, e.g. "treat each other with respect," "communicate directly with people when you have an issue with them," or "have fun."

When worked on at a more specific level, any one of these statements can have numerous and divergent meanings to different people. What one person thinks is respectful treatment, direct contact, or fun is likely to differ significantly from another person's views.

2. *Agreeing to communications procedures may involve changing specific individual actions.* Architects can agree with contractors in that they are willing to have fun, make a profit, or communicate with respect on a project with little threat to the architect's well-worn everyday work habits and methods.

Keeping an accurate, timely log of requests for information, however, may represent a significantly new (and most unpleasant) task for the architect. To a contractor or engineer, keeping an RFI log may be no different than keeping a "punch list" of current tasks and upcoming projects, just one more list. To an architect more used to working with global design, keeping an RFI log may be uncomfortable, unfamiliar, and awkward.

3. *Many communications procedures involve sharing information people may be used to keeping to themselves.* Few people in design and construction would argue against the need for better sharing of information across the disciplines working on a project. Acting on this intention to share information can be a problem, however, because even if just a few people withhold information, that minority can set a tone for the rest of the project.

Simple habit is probably the more serious threat to sharing information, though. Because of the highly fragmented nature of the design and construction industry, people have grown familiar with the need to communicate only a small amount of what they do, and then only to a small number of people. Agreeing in principle to share more information is easy. Remembering to share information when other, more tangible job pressures are mounting is the larger problem.

   4. *To some, communications procedures seem too formal to work.* People with a long history in the design and construction business who like to remember when "we did all the business we could and sealed it with just a handshake" may resist the apparent formality of communications procedures.

   Yet projects are more complex these days, and so are the communications needs that they have. If communications are complex, simply wishing that they return to the way they used to be is not going to accomplish much.

   At the same time, people often find that improved communications procedures work to put some of the fun back into the work. Once the tangled web of project communications is cleared up a bit, it becomes a great deal easier to have trust in one another and derive more satisfaction from the job.

Written Communications

To make the subject clearer, we divide communications procedures into written procedures and procedures for meetings. Written procedures include the request for information process, backup access, and project chain of command.

The Request for Information Procedure

Many of the most serious delays in design and construction projects occur in the RFI loop. One contractor we work with calls RFIs the "Bermuda Triangle" of construction projects: Requests sail into it, and then they are never heard from again. Waiting for a response to an RFI can bring a whole project to a grinding halt, causing delays and overruns.

   Some contractors seem to express outrage at the need to use RFIs at all, but to others RFIs are a natural, inevitable part of design and construction work. RFIs occur at the hand-off point between the architect's design and the contractor's translation of the design into something tangible. Attempting to follow the architect's drawings, the contractor discovers that he or she cannot build what is drawn because of one or more of these problems:

- The drawings are unclear.
- The drawings lack adequate detail.
- The drawings did not anticipate a condition that has evolved since they were completed.
- The drawings specify materials that are not available.

To obtain further clarification from the architect, the contractor submits a RFI on the problem. The function of a request for information is such an integral part of the design and construction process that the American Institute of Architects actually has a standard form for the purpose. See Table 8.1 for a sample RFI log.

Even with such standardization, however, the RFI process often remains problematic and is thus a fruitful area for attention in a partnering workshop. Typical, recurring problems with the RFI procedures include the following:

- The contractor complains that the architect takes too long to respond to RFIs.
- The architect complains that the contractor submits RFIs too easily, never stopping to attempt to figure out the solution on his or her own.
- The contractor complains that the architecture firm "loses" RFIs when they get passed on to members of the firm who may not be working on the project.
- The architect complains that the contractor doesn't fill out the RFI with enough clarity to make it possible to provide a useful response.

Participants in a partnering workshop cannot make these problems go away but they can be creative and effective in establishing procedures to minimize their occurrence and impact. In a partnering workshop, for example, the architect and contractors can

Table 8.1 Sample Request for Information (RFI) Log

Sent by	On (Date, Time)	Topic	Urgency level	Received by	Next step
General contractor	May 3, 10 A.M.	Exterior finish	Medium	Project architect	Forwarded to exteriors vendor May 4, 11:00 A.M.
Electrical contractor	May 4, 1:00 P.M.	Specified fixtures unavailable	High	Project architect	Replace with No. GE 146-981
Plumbing contractor	May 4, 2:00 P.M.	Specified connections violate code	Low until next week, then high	Project architect, then forwarded to plumbing consultant	Call placed to Town Inspector May 7, 11 A.M.

devise several kinds of communications procedures that help RFIs sail through this "Bermuda Triangle":

1. *RFI log.* This is the most useful tool we have seen for managing RFIs. The architect and contractor agree to a matrix format to log all incoming RFIs in the architect's office. There is no set form for this log. We usually find that the architect and contractor can devise a log format that meets their mutual needs in about a half hour if they attempt to chart simply what each needs to know about the progress of RFIs.

On the contractor's side, the form can track when the RFI came in, who sent it, what information is needed, and when the requester would like a response. On the architect's side, it records who received the request, when, whom they may have forwarded it to, and their estimate of a response time. When the RFI is finally completed, the last column, or "RFI complete," is also checked. The architect and contractor together collaborate to keep this log, faxing it back and forth to each other on a regular, often weekly, basis.

2. *Time limits.* Beyond the RFI log, setting hoped-for time limits for responding to RFIs can also help to manage the process. The architect agrees to a general set of guidelines and deadlines for responding to RFIs. If he or she knows a particular RFI will take more time, he or she calls the contractor to let him or her know when to expect a response.

3. *Internal process management.* Many delays in the RFI process are caused by the internal organization of the design firm. The architect hands the RFI off to someone else in the firm who has the technical knowledge to provide a response but whose priorities lie with another project. Or, in large design firms, it is possible to lose track of who passed on the RFI to whom.

Thus, managing RFIs may mean that the project architect has to take a more active role in working within the firm to keep tabs on RFIs. On several occasions, project architects have told us that being involved in partnering made it easier for them to manage their own firms. People in the firm who knew the project was a "partnering project" took extra care to manage RFIs because they knew partnering put the firm under scrutiny.

"We didn't want to be the topic of one of those follow-up workshops," a project architect explained, "so people who are often kind of casual about paperwork and that sort of thing put a little extra effort into turning my RFIs around on that project."

4. *Mutual education.* Architect and contractor can also work together to complete several RFIs in collaboration, providing insight for each other into what makes the RFI effective or not. Discussing a few hypothetical cases in a partnering workshop can save days of frustration back on the job.

5. *Work to eliminate RFIs.* Some people view RFIs as an indication that verbal communications are not effective. In an effort to streamline the communications process overall, they try to set a goal of working towards zero RFIs. In order to achieve this end, they usually have to spell out how verbal communications will work. They also have to work hard at providing the follow through necessary to build the trust that makes such an arrangement work in the real world.

Backup Access and Project Directory

One of the most productive tools that can come out of a partnering workshop is a comprehensive list of every possible means of reaching every project team member: names, titles, work addresses and phone numbers, home addresses and phone numbers, pager numbers, fax numbers, electronic-mail addresses, vacation house numbers, and car phone numbers.

Not too many years ago, it was much more difficult to communicate. The technology simply was not available. Now the challenge is not so much the technology as the manual follow through to make sure everyone has access to the technology. Pagers, electronic-mail, and fax machines lose their value if people who need to have their numbers can't find them when they need to.

Once the list of key people and numbers is drawn up, it is also useful to organize, post, and manage it. The best lists we see fit on one page. The printing can get small, but the ability to find all the necessary information on one page makes the page easier to use overall.

Posting the list in key places (usually in the construction trailer and in the office of each project team member) helps ensure that the information will be available when people need it. In addition, posting the list makes it accessible to people outside the immediate project team who may still need to use it for access of their own.

Of course, making the list available also begs the issue of managing the list, attempting to control the way it is used. This is another issue the project team can take up at the partnering workshop. People who list access information about themselves usually want to make sure the information is not misused in two ways: invading their personal time and invading their privacy. If people provide their home telephone numbers, the project team must also work on rules and procedures to respect the wishes of people who want to be reached only in real emergencies. Similarly, if the list is posted, there must be careful management of who can use it.

People and positions on a project change frequently. Even when the people remain stable, their phone numbers often change. Thus the final aspect of managing the list involves updating it. This is a task that can often be done quickly and efficiently at follow-up partnering workshops.

Clarifying Project Decision Making. Waiting for a decision can cost projects time and money, yet identifying and locating the right person to sign off on some work can be difficult months into a project or if turnover has occurred. Partnering workshops can produce two useful documents in the area of decision making: general principles of project decision making and clarification of the project chain of command.

The general principle most projects try to work with is that of empowerment. In other words, the person closest to the task and the information is empowered to decide as much as possible. This encourages people taking ownership and responsibility for their actions, builds pride, and makes it more possible for two people to resolve differences on their own without help. Manufacturing businesses use this principle in many situations.

In order to make the principle of empowerment work, however, it is also important to articulate the boundaries of the individual and the chain of command for making decisions. A typical chart summarizing this information would include a listing of each project team member along with the kinds of decisions he or she is empowered to make: scheduling, budget, changes, authorization for expenditures, etc.

First Person: Participant Comments about Communications Procedures

With all my experience in this business, over 20 years, I can't believe I never saw this before. It seems like such an obvious thing that you have to manage communications if you want to get it right.

General Contractor

When we were working on all the meetings procedures, I thought, "This is a waste of time." Once we finished, though, I could see that by spelling things out, we're going to make all of our meetings a lot more worthwhile.

Electrical Contractor

I like to think of myself as a pretty good communicator, but working on all these procedures reminded me of how much I often leave out. It was kind of painful to work on, but I can see that the success of the project depends as much on my ability to communicate the design as on my ability to conceptualize the design.

Project Architect

I've been involved in design work before when people tried to be nice to me, but they never really paid attention. Now that we have clear procedures for meetings, I can see that it will be a lot easier for me to represent my concerns and be heard by the project team.

Building Manager

9
Conflict Resolution Process

Summary

Conflict resolution in partnering consists of both paperwork (procedures and steps to follow) and behavior (everyday actions and attitudes). Both can be deceiving. The paperwork can appear meaningless, a cut-and-dried set of steps that seem obvious if not boring. The behavior can look frightening, a set of individual actions that encourages people to resolve conflicts in ways that may not be familiar to them.

We think otherwise. Effective conflict resolution includes three different kinds of work that are essential to partnering's overall effectiveness:

1. *Designing strategies.* This involves planning, imagining, and thinking through in a careful, thoughtful way the areas of the project in which conflicts and disagreements are most likely to occur.

2. *Working with empowerment principles.* In conflict resolution, empowerment means developing procedures that empower people to solve their own conflicts.

3. *Writing procedures.* Writing, with the involvement and approval of the full project team, very specific steps that articulate how the people who have the disagreement are to proceed.

Effective conflict resolution also depends very much on the ability of project team members to communicate in a collaborative way. It is very helpful for conflict resolution when project team members develop an attitude of comfort with conflict and develop confidence that conflict is not necessarily negative but inevitable. Also advantageous is the feeling that conflict can, in fact, be resolved by following the steps that the group develops.

We explore the kinds of one-on-one communications skills that contribute to effective conflict resolution, later in Chap. 11, subtitled Improving One-on-One Communications Skills.

Perspective on Conflict Resolution

Conflict resolution overall has a bad reputation. It represents, to some, the aspect of partnering with the least potential and also the aspect that is the most reactive, the most negative, and the part that is necessary to use only if the other (more important) parts of partnering fail. To others conflict resolution represents the only important part of partnering, the part that spells out what to do when things go wrong, as "we know they will."

An effective conflict resolution process is none of these (see Table 9.1). Rather, conflict resolution is an integral part of the documents and procedures of partnering, the third leg of the three-legged stool that also includes the goals statement and communications procedures. Conflict resolution is not negative at all but quite positive in that it views conflict as a potentially positive force if it can be managed. Trying to manage conflict that is likely to occur is not pessimistic or optimistic but realistic.

Conflict resolution is not reactive at all, but quite proactive and strategic in that it asks partnering participants to anticipate the ways in which conflict is likely to occur on the project and to design in a careful, thoughtful way strategies to handle conflict when it does arise.

Although it may not be the most important aspect of partnering, conflict resolution is often the most visible. Goals and communications procedures may determine the parameters within which conflict resolution must work, but the conflict resolution procedures are the predominant aspects of partnering one notices when real conflicts erupt.

Table 9.1 Misconceptions and Realities of Conflict Resolution in Partnering

Misconception	Reality
Conflict Resolution	
Is reactive	By anticipating problems, is proactive
Works within a "negative" mindset	Works within the "positive" mindset that people can often resolve their own conflicts
Is what people do only when partnering fails	Is an integral part of partnering along with goals statements, communications procedures, and skills training
Is "just more paperwork"	Provides necessary clarity of steps to escalate conflict slowly and thoughtfully if at all
Means I have to learn to live with compromise, taking second best for myself and the project	Means I have to learn skills to work through conflicts in collaboration that provide improved results for both parties and for the project overall

The most common misconception people hold about conflict resolution, though, is not with the procedures and paperwork but with what it implies for settling the real conflicts that arise on a job. To many conflict resolution means giving up, knuckling under to compromises so that they are unable to get what's best for them or what's best or the project. Conflict resolution actually means acquiring new skills for communications so that people put their efforts into creative problem solving instead of into defending their positions. When they can do this, they do not have to settle for compromises at all and the project benefits overall.

Designing Strategies

With just 30 minutes to go in a recent two-day partnering workshop, we asked the project team, "Are you sure you've thought of all the things coming up in the next few months that could cause any kind of miscommunication or conflicts?" The twelve people seated around the table looked up with the understandable frustration of people who had already been asked the same question numerous times over the course of the past few days and again in the past hour. Also, it was the end of the day and at least a few members of the group were beginning to think more about how a few extra minutes lead time on the road now might help them save many more minutes in their commute home later. Nonetheless, we held the uncomfortable silence for a few more minutes, just to be certain.

When no one ventured any additional concerns, we acknowledged that we had pressed as hard as possible and began to hand out the feedback forms to evaluate the workshop. As the papers circulated, most members of the project team began to pack up their belongings and complete the forms.

One of the administrators in the client organization hesitated for a few moments, though, staring out the window. Finally, she asked tentatively, "Have we thought about the training?"

"Of course we have," her boss cut her off, snapping his briefcase shut and looking impatiently at the rest of the group. "We always provide ample training for our managers."

"No, that's not what I mean," the administrator continued. "I was thinking more about the training that's going to be necessary for the boiler operator, the building electronics people, and the telecommunications workers."

"Are you crazy?" her boss interrupted, growing angry. "That's months and months away. We have plenty of time to worry about that when the time comes. Can't we get out of here early for just one time?"

"I really don't want to cause any inconvenience; this will just take a minute," the administrator continued, cautiously choosing her words. "We had that project in the southeast region last quarter. The whole building was complete, done up real nice, even a little under budget until that point. Unfortunately, occupancy was delayed for over a month, almost two months, because they had to wait for the technical training crews to come through for the building staff."

"She may have a point," the general contractor added. "I've seen it on a few other buildings lately that are using the same kinds of technical systems we're using here. There's not enough people certified to provide the training, so they're booked very tightly. You've got to schedule the training months ahead of time if you want to sequence it into the building opening date."

"From the way you're saying it," the client's senior manager replied, "I take it you mean we should do this now." Without even waiting for the contractor's answer, he reopened his briefcase and began to make himself more comfortable for another half hour's discussion.

Although we can know that the design and construction industry is a reactive business, our knowledge does not always help to avert avoidable conflicts. Nearly every partnering workshop we conduct involves moments such as those noted above, in which people struggle with great difficulty to remove their attention from tomorrow's problems so that they can think wisely about the problems lurking two, three, five, or six months down the road. Effective conflict resolution begins with strategic thinking about those problems.

Why Strategic Thinking Is Difficult

Few people disagree about the need to engage in such strategic thinking, but most find it difficult to do, and for good reasons.

New Projects Breed Optimism. A real construction manager provided the quote in the State Records Building Financial Services case later in Chap. 15: this text, "At the beginning of every project, it's as if everybody is in love with everybody else. It's as if no one ever built anything before; it's as if no one ever had a disagreement."

The project-based nature of design and construction makes it possible for people to view every project as a real new beginning with a whole new set of possibilities. That helps people approach the new project in a fresh manner, but it may also lead people to be more optimistic than their own experience warrants.

Short-Term Problems Seem Pressing and Cloud Long-Term Vision. By the time projects are a few months underway, there is usually a backlog of real problems that must be solved in the short term. To fully attend to more long-term issues at that point is a challenge and may even feel irresponsible.

Long-Term Planning Is Abstract. Given a choice between working on a short-term problem with a small influence on the project and a long-term issue with a great influence on the project, project team members consistently choose the short-term one because the long-term issues seem abstract. Even when one succeeds at resolving them, there is no immediate, tangible product that acknowledges or marks success.

The Industry Surrounds Individuals with a Short-Term Focus. To develop a more long-term focus means, for individual project team members, working against an industry that is driven by a crisis or by short-term orientation. It would be difficult to change individual behavior to a more long-term focus under any circumstances, but it is extremely hard when the nature of the business keeps generating crises on an hourly schedule.

The Industry Favors Action over Thinking. Strategic thinking begins with intelligent reasoning about the problem. Action steps, if any, come from the thinking. Unfortunately, the kinds of introspection and reflection that are the hallmarks of thoughtful problem solving have not always been strengths of the design and construction business (see Table 9.2).

Specific Strategic Thinking Tasks

To address many of the issues above, we follow a three-part sequence that begins with asking workshop participants questions, then asks people to think about the causes or sources of problems behind their answers, and finally lists possible action steps. This sequence helps direct their thoughts away from the present into the future while also helping to keep them focused on describing and solving the problem rather than leaping to conclusions and premature action steps (see Table 9.3).

Because this activity is a bit of a fishing expedition, it is useful to keep it relatively loose. In terms of the questions, this means not

Table 9.2 Why Strategic Thinking Is Difficult in Design and Construction

- New projects breed optimism.
- Short-term problems seem pressing and cloud long-term thinking.
- Long-term planning is abstract.
- The industry surrounds individuals with a short-term focus.
- The industry favors action over thinking.

Table 9.3 Strategic Thinking for Design and Construction
A Template for Anticipating and Averting Conflicts

Strategic Questions	Responses	Reasons	Action steps
What are three issues coming up on the project that you think people will disagree about?			
If it were three, six, and nine months from today, what items would be on your "to do" list?			
What conflicts came up on several of the past projects you worked on?			
What aspects of this project will take more than the usual amount of communication?			

working too hard at trying to find the "correct" question but instead being patient with asking several questions. As long as the facilitator is attentive, the group's energy and interests will lead close to the key issues no matter what the starting place is. Specific questions useful to help people design strategies about conflict resolution include:

- If it were six months from today, what key items would be on your "to do" list?

- Which long-term project items require that we get started working on them today?

- What issues did people not address soon enough on the last three projects you worked on?

- What are three things coming up in the project (in three months, six months, and a year) that you think people will disagree about?

- What are three areas of the project in which it will take more than the usual amount of communication in order to ensure coordinated action?

- In your last three projects, what arguments did you find yourself involved in?

Working with Empowerment Principles

Several principles of everyday behavior underlie an effective conflict management process. We call these *empowerment principles* because their chief function is to empower people to resolve their own conflicts and disagreements. Often, participants in partnering workshops develop and formally sign off on a similar list of principles to provide the guidelines for conflict resolution on the project.

Empowerment principles begin from two different places: the individuals involved in the disagreement and the organizational structures and processes of the firms involved in the project (see Table 9.4). For individuals, empowerment principles include the following:

1. *Speed.* People on the project will try to resolve any disagreements with one another quickly. The more time passes, the more difficult it is to resolve an issue. Informally agreeing to never leave the job site until one has addressed the issue is one way to make this principle more specific.

2. *Initiative and responsibility.* Bruised egos and hurt feelings can get in the way of resolving conflicts. When people have a disagreement or take issue, it is their responsibility to take the initiative to seek out the other person in an attempt to resolve the problem.

Table 9.4 Empowerment Principles for Conflict
Resolution

Individual	Organizational
Speed: People try to resolve issues quickly	*Autonomy:* Keep decision making down as low as possible in the organization
Initiative: Individuals take responsibility for resolving their own conflicts	*Clarify:* Clarify chain of command and decision-making authority
Directness—no messengers: People deal directly with the people with whom they have conflicts, not work through others indirectly (as messengers)	*Coach:* Help people resolve conflicts but do not take the responsibility for resolving it away from them
Spirit of cooperation: Entering into a conflict, people maintain a spirit of cooperating with one another	*Model collaboration:* People in the organization act in congruence with the collaborative approach on the job
Try to resolve before escalating: Before taking an issue to their bosses, people make a serious effort to resolve it with one another	*Support:* The organizations reward individual empowerment

3. *Directness.* People agree to go directly to the person with whom they have the disagreement rather than escalating the issue immediately or carrying it to a third party to "vent." People usually have no trouble agreeing to the directness principle in theory, but they often encounter difficulties in living up to it.

   a. *No messengers.* This corollary of the directness principle addresses the most common violation of directness, i.e., people complain to a third party, expecting that the third party will carry the message on their behalf. (Sometimes the third party is a supervisor or manager.)

   Although most people in partnering workshops readily admit that they know that the fate of messengers typically is to "get shot," people persist in recreating the problem. Supervisors and managers in particular sometimes feel that it is their responsibility to solve problems that their people bring to them. The underlying problem here is that it is usually not possible for anyone outside the immediate situation to devise a lasting solution to problems between other people.

They and others on the job site need to learn, instead of trying to solve the problem directly, the following coaching behaviors:

- Helping the person think through how to handle the problem and encouraging the person to handle the problem independently.
- Bringing the people together face to face and mediating their conflict, still helping them to find their own solutions.

4. *Spirit of cooperation.* In addressing a disagreement, it is up to both people to work in a spirit of cooperation. That means approaching the other person with the attitude that "there is a disagreement, let's work on trying to resolve it."

5. *Try to resolve before escalating.* For individuals, the first step in a clear chain of command is to meet with the other person for a designated period of time, working on the problem, before escalating it.

Often, participants at a partnering workshop devise the procedure that people who have a disagreement meet once for at least an hour and a half, then meet again for another hour and a half before bringing the issue to their bosses. These meetings may take place on the job site or on neutral territory such as a diner. After that, the people involved may both meet with their bosses present before escalating the argument.

For organizations, empowerment principles include the following:

1. *Keep decision-making authority as low in the organizations as possible.* This encourages people to take responsibility for solving their own problems while also placing decision-making power in the hands of the people who have the best information about an issue: the people most directly involved with it.

2. *Clarify the chain of authority and decision making.* In order for people to make effective decisions, they must have enough information to know what they can and cannot do. They must know what the next step is in case the step on which they are working falls through.

3. *Provide help to people involved in a conflict but do not take away the responsibility for solving the problem.* This amounts to an organizational policy of "no messengers." When bosses step in to settle a dispute, they get short-term results at the expense of long-term reluctance to deal directly with problems ever again.

4. *Model a collaborative style of communications.* If the senior managers of the firms involved in partnering use a highly confrontive approach to resolve conflicts, it is inevitable that the model they create will ripple throughout the project.

5. *Actively support individual empowerment.* One of the easiest things for an organization to do is to write memos proclaiming support of individual conflict resolution methods such as those described above. Using those methods is another matter, though. If organizations want to support empowerment in conflict resolution, it is helpful for them to recognize and reward such behavior formally in performance discussions and informally in praise, recognition, and encouragement from managers.

Writing Conflict Resolution Procedures

The process for writing conflict management procedures is the same as that for writing goals and communications procedures, but with a different starting point. While goals and occasionally communications procedures come more directly from workshop participants' hopes and experiences, conflict resolution procedures come primarily from the group's work described above in analyzing the likely problems the job will encounter, designing strategies for resolving the problems, and devising empowerment principles.

The facilitator breaks the group into subgroups (to make it more possible for people to work on the issue in depth), refers to the work the group has done in developing strategies and empowerment principles, and asks each subgroup to prepare three or four conflict resolution procedures to propose to the overall group.

Prior to breaking the group into subgroups, the facilitator may also ask everyone to take a moment or two in silence and jot down one or two possible procedures. Writing ideas down in such a manner helps people who are not so comfortable working in groups to clarify their thoughts before entering into the group setting.

Under these conditions, the subgroups will usually cover the major issues useful to the project. To further guide the participants, the facilitator may ask the overall group, prior to the division into subgroups, what kinds of issues the procedures should address, i.e., what issues people are likely to disagree on (see Table 9.5). This discussion should go quickly and lead to fairly complete answers if the group has done its work with designing strategies.

Typical issues include the following:

- *Scheduling conflicts.* Project schedule changes are inevitable and, because they can cause ripple effects for numerous firms working on the project, can lead to conflicts about sequencing the work. Effective conflict resolution procedures spell out what happens when delays occur:

  Who takes the lead in rescheduling

  What factors are considered and weighed

  Compensation for losses due to rescheduling

  Who participates in rescheduling discussions and with how much authority

- *Use and management of requests for information (RFIs).* Contractors, subcontractors, and architects on a project may have very different experience in working with and very different expectations for the use of RFIs. (The State Records Building Case described in Chap. 15 illustrates how this problem evolves and is resolved on a real project.

Table 9.5 Typical Project Issues to Anticipate and Address in Conflict Resolution Procedures

Scheduling and coordination problems

Request for information use and management

Contact points: who talks to whom, when, and about what

Spending authorization and procedures

If problems and conflicts arise, focus not on blame but on
- Analyzing why
- Designing thoughtful action step repairs

Currently, architects are in conflict with contractors on many projects because they think the contractors use the RFIs for small issues that could be cleared up with a phone call. On the other side, the contractors say they need to document more these days just in case the schedule slips, so they can show a paper trail of their own actions.

- *Contact points: who talks to whom.* A project's communications procedures should map out who will pass on what information to whom and when, over the course of the project. Working on conflict resolution, the issue in this area has to do with what happens when people don't pass on the information they were supposed to.

- *Spending authorization and procedures.* One of the recurring areas in which conflict occurs is related to spending money. Problems can arise in two areas:

  People make mistakes about what they themselves can spend or authorize.

  People do not know who on the project can authorize what level of spending, and waiting to find out or processing payment forms incorrectly leads to delays.

Conflict resolution procedures for any of these items cover two major areas:

- *Determining why the problem occurred.* This may take some effort, discussion, and thinking. It may also be difficult to allot time for this determination when other more immediate problems face the project. Nonetheless, if people do not fully understand why the problem developed, they will probably not be effective in preventing it from happening again.

- *Designing and implementing clear actions so that the problem does not occur again.* Even on projects of less than a year's duration, conflicts have a way of recurring because people fail to take the time to unravel why they occurred and fail to take appropriate action to avert their recurrence. Usually it takes just a few hours to analyze the problem and devise new action steps so that future conflicts can be avoided. This effort amounts in many cases to adding communications procedures and goals to the original set of partnering documents.

There is a temptation in conflict resolution for the parties impacted by a problem to leap immediately to focus on fixing blame and demanding compensation. This is understandable but, even in what may seem to be clear-cut cases, it is expensive, unpredictable, and largely unworkable.

Furthermore, when people focus more on understanding the problem and devising clear action steps for future situations, the need for blame and compensation diminishes. The more clearly people understand the problem, the less likely they are to attempt to assign blame when blame is not clear. The more people focus on future actions, the more satisfied they become that compensation for past problems is not necessary.

First Person: Conflict Resolution

I didn't think it would be necessary to have conflict resolution procedures in place after all the work we did on goals and communications. I was wrong; the few conflicts where we used the procedures more paid back many times for the time we invested in writing them.

Project Architect

I thought one of the most useful things we did was to write these procedures. Just writing them helped clear the air and will probably help us to avert many of the conflicts that would have occurred.

HVAC Engineer

It seemed strange to me that we would have to work on conflict resolution after all the other work we did; it seemed to take away from the other partnering tasks. Once we started working on conflict resolution, though, I could see it was much more proactive than I thought it would be.

General Contractor

We like to think that we are so unique in this field, that every building is new. I thought it was especially interesting in conflict resolution that we were able to so quickly predict the areas where we knew we would argue.

Corporate Real Estate Manager

3

Training and Infrastructure

This part contains material essential to improving partnering effectiveness. Beyond the partnering documents, it is important to provide project team members with skills and a project infrastructure to implement the documents. It is probably even more important to ensure that project team members have the skills and structure on a day-to-day basis, to carry out the intentions partnering has of strengthening working relationships.

Two chapters (10 and 11) in this part describe training that supports partnering, and Chaps. 12 and 13 describe project infrastructure to support improved communications.

10
Valuing Differences

Summary

Projects ask people to work together as a team. Few people would argue with that expectation. However, people on a project seldom have the time or the opportunity to learn enough about one another to make the teamwork goal attainable.

Working with a personality profile can accelerate project team members' understanding of each other while also providing information that can prove very useful for later efforts in conflict resolution and overall project management.

This chapter examines why using personality profiles can improve partnering effectiveness. The chapter explores in detail the specific uses of the most widely used personality profile, the Myers–Briggs Type Indicator, for design and construction communications and Partnering.

Using Personality Profiles

Numerous personality profiles and tests of individual behavior are available to group facilitators: the Fundamental Interpersonal Relationship Orientation (FIRO) and FIRO-B, the "What's My Style?", the Neurolinguistic Communications Profile, the Interpersonal Influence Inventory, the Learning Style Inventory, the Selling Skills Inventory, etc.

Table 10.1 How Working with Personality Profiles Improves Project Team Communications

1. Minimize blame.
2. Induce individual reflection.
3. Improve understanding of others.
4. Provide a "quick read" of new people.
5. Provide new insight into long-standing relationships.
6. Explain much conflict...and chemistry.
7. Enable people to translate ideas so others can understand them.
8. Help people manage everyday conflicts.
9. Provide a road map of likely group communications problems and suggest action plans to address them.

These instruments are popular training and development tools. They are especially valuable in partnering because they enhance individual learning and the quality of group discussion (see Table 10.1):

1. *Minimize blame.* When people work in a group or team they can see clearly how other people in the group create problems, but they can be blind to what they themselves do to make matters worse. Working with personality tests gets people to think about themselves and their impact on communications, the group, and the problem. Thus the instruments often help individuals shift their individual stance from blaming others to wondering how they themselves are making the problem worse.

2. *Induce reflection.* Once people have stopped focusing on others, they begin to reflect in greater depth and detail on themselves: Why do I avoid certain kinds of communications? How do I make decisions? Do I ask for too much detail in trying to solve problems? Do I arrive at judgments too quickly? Such reflective questions do not have a correct answer, of course. They get individuals to review the habits they bring to the job and identify behavior that may not always be productive.

3. *Improve understanding of others.* When people do not understand one anothers' reasoning, they often become frustrated, even suspicious. They may think that the other person is doing things

intentionally to upset them. Improving one's understanding of others increases the level of trust among individuals.

4. *Provide a "quick read" of new people.* People who don't know one another well if at all are often thrown together on a project team and expected to function as if they had been working together for years. Working with personality profiles enables the project team members to get below the surface and discuss at a deeper level what makes them tick, what motivates them, how they make decisions, etc. This kind of discussion accelerates team building because it enables people to discuss aspects of their behavior that impact group communications.

5. *Provide new insight into long-standing relationships.* Often, members of a group of people who have worked together for some years will comment to each other, after working with a personality profile, "I finally understand you." The pressures and tensions of everyday work do not often make it possible for people to discuss their underlying motivations and preferences. Working with a personality profile helps people who have worked together for some time to understand one anothers' reasoning better.

6. *Explain much project conflict…and chemistry.* Much of the conflict that arises on a project is rooted in "bad chemistry" between people. For example, the contractor's on-site superintendent and the project architect just don't get along. The contractor thinks the architect is sloppy, abstract, and arrogant. The architect thinks the superintendent is picky, negative, and out to make money on unnecessary change orders.

Some of these allegations may be true, but what is also true is that both persons are different types of people with different preferences. The architect literally does not see details. She is not at all lazy, just not interested in the same issues that interest the superintendent. She is indeed abstract but not intentionally; that is just the way she sees things. She does not mean to be arrogant, that is just the way others perceive her.

Likewise the superintendent is not picky but just very concerned about job issues he gets paid for. Nor is he negative; he is just realistic based on his own past experience. As far as exploiting change orders for profit, he is really just trying to cover costs.

The "bad chemistry" between the superintendent and the project architect seems like irreversible conflicts, all rooted in legitimate arguments. It is also possible to turn an understanding of these conflicts around to focus on preferences. Bad chemistry here is really the result of differences in the architect's and the superintendent's preferences, biases, and communications styles.

Conversely, some of the "good chemistry" that occurs on a project—i.e., two or three people who "just seem to hit it off"—is often the result of shared preferences and biases.

7. *Enable people to translate their ideas into a language the other person can understand.* Many project communications problems arise not because people disagree but because they describe the same issue in two different languages. Once people know each others' communications styles, they can translate their ideas into a wording and language that the other person can understand.

8. *Help people manage everyday conflicts.* When people who have worked together for many years work with personality profiles, they often comment to one another, "I finally understand you....You still drive me crazy, but at least now, I understand why."

It is a bit more difficult for people to lock horns and remain locked when they understand each other's personality a bit more. The superintendent can remind himself that the project architect is not trying to start a fight, she really does think there is enough detail in the drawing. The architect can remind herself that the contractor is not trying upset her, he really does need more detail in the drawing in order to understand it.

9. *Provide a road map of likely group communications problems and suggest action plans.* Beyond what individuals get from working with personality profiles, the whole project team can gain perspective into likely pitfalls, problems, and obstacles it will confront. Anticipating these problems, the group can take action to improve its communications.

For example, if most of the members of the group prefer to work on their own and dislike meetings, the odds are the group will not meet often or long enough. If most of the members of the group are highly opinionated (a likely condition in design and construction), then it is likely that they will argue more than necessary and find it difficult to change their decisions.

Knowing these tendencies, the group can step back from troublesome decisions and review their solutions in light of their preferences and biases. Groups of people who don't like meetings may discipline themselves to meet more often, knowing they tend to avoid meetings and might miss important information.

The Myers–Briggs Type Indicator

Of the personality profiles available for partnering, the Myers–Briggs Type Indicator (MBTI) is among the most useful. Selling over three million copies a year for applications ranging from management training to career counseling to team building to marriage counseling, the MBTI is widely accepted.

Isabel Briggs-Myers and her mother, Katherine Briggs, devised the instrument in the 1940s in an effort to describe with a test different dimensions of character and personality. Then Karl Jung also identified these dimensions: introvert–extrovert, intuition–sensing, thinking–feeling, perceiving–judging. Since their original efforts, others have worked with the basic concepts so that at present seven different versions of the test exist, including versions in different languages and a version for children.

Having used many personality profiles, we find the MBTI:

- Provides extensive insight for the amount of time invested in working with it
- Is complex enough to avoid jargon and describe real project communications issues
- Is simple enough that laypersons can understand and use it after just a few hours of training
- Is both credible and accessible to a wide range of audiences, with an extensive research base and many books and programs using it
- Is less vulnerable to being misused than other instruments.
- Provides information that is very useful for partnering because it focuses on peoples' communications styles

Using the Myers-Briggs Type Indicator in Partnering: Parameters

The MBTI gives respondents information on four dimensions of their personality: introvert–extrovert, intuition–sensing, thinking–feeling, and perceiving–judging. In all of these it is essential to keep in mind several factors to guide the optimum use of the instrument (see Table 10.2).

- *Each dimension is not an "either/or" trait but a continuum.* A person is not either an extrovert or an introvert, for example, but someone who has a preference for either trait. The preference may be moderate, strong, or weak. Even if the preference is strong, people still function somewhat on the opposite side as well. Thus a person with a strong extrovert preference still operates at some times through his or her introverted side.

- *In communications as well as in adult development, the challenge is to understand and appreciate one's opposite or "shadow."* Understanding,

Table 10.2   Guidelines for Using the Myers–Briggs Type Indicator in Partnering

---

Each dimension is not an "either/or" trait but a continuum. (People are not "introverts," rather, they have an introvert preference of some level.)

The challenge in adult development as well as in communications is to understand and appreciate one's "shadow," i.e., one's opposites.

The instrument provides only approximately accurate results in public workshops.

Participants should use information about type as a means of *explaining* project issues and problems, never as

- Evidence ("You intuitive types are all alike.")
- Excuse ("I'm intuitive; there's enough detail in that drawing for me.")

Use type information as a means to the end of project improvement, to explain:

- Recurring arguments
- Recurring interpersonal knots
- Miscommunications and misunderstandings

Use type information as context to address subjective questions and problems.

coming to terms with, and appreciating one's "shadow" or undeveloped side is a major theme in adult development. Communicating with people outside one's preferences likewise involves inordinate confusion and energy. In both instances, it is extremely valuable to work with the "shadow" in order to incorporate completeness in the final product.

- *The instrument provides only approximately accurate results in public workshops.* The MBTI asks respondents to report on deepseated beliefs, assumptions, preferences, and biases. The surroundings and applications of the instrument influence what people report as their preferences. Usually people in a partnering workshop are not overly concerned about the uses (and possible misuses) of their scores, so they report preferences fairly accurately. Nonetheless it is essential to ask people how accurately their scores reflect their real preferences.

- *The MBTI should not be used as an excuse ("I won't come to that meeting, I'm an introvert) or as evidence ("You 'thinkers' are all too cold and insensitive, that's why there is so much conflict on this project") but as a possible explainer.* For example, what do our preferences say about:

  What we argue over

  Who (apparently) wins

  Why we miscommunicate frequently over a certain topic

  Why certain people on the project tangle with each other

  Information about type is always a means to the end of project improvement, never an end in itself.

- *A second valuable use of personality profile information is in addressing subjective questions and conflicts.* When the architect knows the contractor is "sensing" (detail oriented) and the contractor knows the architect is intuitive (abstract, thinking about the big picture), they can both then get down to business addressing the difficult question: What level of detail in drawings *would be optimum for the project?*

Using the MBTI for Partnering: Specifics

We have administered various versions of the Myers–Briggs Type Indicator to thousands of architects, engineers, contractors, and corporate real estate managers in seminars and consulting assignments. These are some of the applications for partnering, taken one aspect at a time. The MBTI gives respondents information on four dimensions: introvert–extrovert, intuition–sensing, thinking–feeling, and perceiving–judging.

Introvert–Extrovert

According to several sources, extroverts outnumber introverts in the United States by a factor of three to one. This percentage is probably reversed in some Eastern cultures.

Many people mistakenly think that introversion implies shyness. That may sometimes be the case, but it is not the root of the preference. More accurately, people with an introvert preference are energized internally by their own thoughts and feelings, while people with an extrovert preference are energized by things around them. Extroverts are energized by interaction with others; introverts are drained. With his visible enjoyment of the interview process, President Bill Clinton is probably a strong extrovert, for example. Former President Nixon, on the other hand, would probably have scored as more introverted.

It is usually easier to understand extroverts more quickly because they talk about themselves more and demonstrate more expressiveness on their faces and in their tone of voice. People misread introverts more often because introverts introvert their deepest thoughts and feelings, i.e., direct their thoughts and feelings inward.

Operating outside one's preferences often feels awkward; it does not seem as if one is accomplishing anything. Thus extroverts criticize introverts for not sharing enough information and also for not doing it in a timely way. Often introverts do not see the need to share information.

An extroverted corporate real estate executive was dismayed to discover at a partnering meeting that his company's project liaison was not maintaining daily phone contact with the construction site. The executive's anger surprised the site supervisor. An introvert herself, she didn't see the need for the liaison to call every day, even though the two of them had a backlog of miscommunications rooted in their infrequent communications.

On the other side of the coin, introverts criticize extroverts for sharing too much information too often, for meeting with no purpose, and for invading their space.

Extroverts and introverts on a construction project also may tangle over the meaning of what they are saying. Extroverts tend to talk and interact as a way of thinking, "talking through" an idea as a way of developing their thinking. In some ways, extroverts do their best thinking when they are talking.

Miscommunications arise when introverts assume that what the extrovert is saying represents a conclusion rather than just an ongoing process of idea formulation and development. Introverts instruct extroverts, "Don't come to me with problems, come with solutions."

Conversely, when introverts express an idea, it is usually only after they have thought about the idea for some time and reached a conclusion. Thus when introverts talk they are less likely to change their minds quickly when presented with new information. When put on the spot with a demand to provide a quick reaction to a new idea, the introvert is likely to respond, "I'll get back to you."

See Table 10.3 for a listing of communications issues related to introvert–extrovert traits.

Intuition–Sensing

This aspect of the MBTI has extensive applications for partnering. It explains many of the recurring, classic conflicts and miscommunications that arise on projects (see Table 10.4).

The sensing and intuition function begins with what people see. Looking at a cluster of trees, sensing types see what's really there, noting each tree in detail, right down to the number of leaves. Intuitive people see possibilities, so they look at the same trees and see tables, newspaper, a song, or a dream.

Only about 25 percent of the U.S. population has an intuitive preference, but many more architects in our seminars, often as many as 75 or 80 percent, score as intuitive types. The reason for this occupational clustering may be that architecture is one of few professions that rewards intuitive thinking. Many of the profes-

Table 10.3 Project Communications Issues Resulting from Introvert–Extrovert Preferences

Introverts may:	Extroverts may:
■ Not share enough information in a timely way	■ Share too much information
■ Fail to see the need or point in sharing information or in meeting	■ Meet too often, for too long
	■ Appear careless in talking through an idea
■ Get misread	■ Invade introverts with unnecessary requests and conversation
■ Think extroverts are committed to an idea they are merely vocalizing in order to "think through"	■ Think introverts are more flexible when they have already decided on something
■ Not participate adequately in project meetings	■ Inadvertently dominate project meetings

■ Two introverts working together may not meet or communicate often enough to exchange necessary information.

■ Project teams that have a significant number of introverts may have recurring information gaps.

■ Introvert–extrovert differences among people who work together often cause friction.

sional trademarks and icons of the design profession reflect the intuitive preference:

■ The motto of some graduate schools of design is that "All great architecture leaks."

■ The embroidered wall hanging at one nationally known firm reads "Keep it fuzzy."

■ The fact that for many decades, most architectural awards were given to buildings not yet built.

■ The predilection of architects is to expand on whatever assignment or task the client gives them.

■ Architects and architecture firms struggle to a great extent with managing the detailed (sensing) side of the design process—tracking expenses, logging requests for information, keeping records, or documenting anything.

Table 10.4 Project Communications Issues Resulting from Sensing–Intuition Preferences

Sensing types may:	Intuitive types may:
■ Request an unnecessary level of detail	■ Communicate at high levels of abstraction
■ Document unnecessarily	■ Fail to see the need for detail and/or documentation
■ Resist change that is needed	
■ Not understand an architect's intuitive descriptions and solutions	■ Make changes for the sake of changing
	■ Project unrealistic optimism
■ Project negativity or skepticism	■ Provide unclear directions
■ Perform jobs one at a time instead of tackling multiple projects	■ Have difficulty sticking with one task
	■ Have difficulty completing tasks
■ Have difficulty delegating	■ Not be able to explain in sensing terms what they visualize
■ Not see what intuitive types visualize clearly	

- Many architects are intuitive; many engineers and contractors are sensing.
- Sensing and intuitive types typically argue over the level of detail, following procedures, the need for documentation, the accuracy of facts, and the chain of command.
- Project teams that have a significant number of sensors often have strong boundaries and barriers among the trades and professions.
- Intuition–sensing differences among people who work together often cause friction.

- Most architects are unwilling to repeat a design until they perfect it. Intuitive types usually prefer to move on to something new.
- Most architects work on several tasks at one time with great ease...and struggle to finish any one of them.
- Architects have a genuine ability to look at a set of drawings and "know" what it will be like to live in the building.
- A lack of detail exists in many architects' drawings.

More engineers, contractors (note: we have noticed a significant grouping, though not quite a majority of general contractors, who land more on the intuitive side), and subcontractors register a sensing preference. These contingents reflect the sensing preference in:

- Their preference for facts over concepts.
- Their natural attention to detail.
- Their understanding of the role and impact of the details and fixtures in contributing to the overall design.
- Their discomfort with abstraction and possibilities.
- Their disbelief that architects can really visualize what they describe.
- Their ability to discern quickly in what specific ways a design will not work.
- Their discomfort with working on several tasks at once and their preference for completing the job at hand before moving on.
- The fact that their cars are usually tidier than the architects' vehicles.
- Their discomfort with change, especially change they cannot visualize or understand.
- Their maxims, "A place for everything and everything in its place," "Any job worth doing is worth doing well," "Always finish what you start," and "It's the details and fixtures that make the job."
- Their ability to control and dominate an argument because of their recall of facts and their ability to arrange those facts sequentially.

When workers with opposite preferences meet head-on on a construction job, the result is a whole collection of classic conflicts. Of course not every architect is intuitive nor is every contractor and engineer the sensing type. However, the intuitive–sensing split is divided often enough along occupational lines to make for recurring communications issues:

- Contractor to architect, "There's not enough detail in that drawing." Architect's response, "Sure there is. Ask anybody in my office."

- Architect to contractor, "You're acting paranoid. You don't need to write an RFI for every little change." Contractor's response, "I need to document every detail."

- Architect to contractor, "That's implied from the rest of the drawing." Contractor's response, "If I can't see it, how am I supposed to read it?"

- Architect to owner, "Sure it leaks, but isn't the flow beautiful? Doesn't it make a great statement?" Owner's response, "I think I can make a better statement: We're holding back payment."

Thinking–Feeling

This aspect of type, in which the U.S. population is split about 50–50 (a slightly higher percentage of men are thinkers, a slightly higher percentage of women are feelers), describes the mechanism by which people make decisions. Thinkers prefer to analyze, using some kind of data, and decide based on that analysis. Feeling types decide based more on a "gut feeling."

Feeling types may experience and often project more emotion. Thinkers tend to project a more even level of emotion. It is usually important for thinkers to use logic. It is usually more important for feeling types to use caring and show concern. Feeling types may accuse thinkers of insensitivity and coldness. Thinkers accuse feeling types of being ruled by their emotions.

This aspect of type does not divide so clearly along trade or professional lines in design and construction. The mix of thinking and feeling types on a specific job, however, strongly influences the nature of its overall tone of communications (see Table 10.5).

Thinkers' more typical curt, to-the-point communications style can create a climate of minor conflicts and brush wars on a job site. Thinkers usually don't mind conflict, sometimes even enjoy it. Feeling types are more bothered by conflict and so may avoid it or procrastinate dealing with it.

Thinkers criticize easily, seldom praise, and analyze others' pain instead of providing support. Thinkers may feel awkward working on the feeling side. The Lilith character on the old "Cheers" television series illustrates this phenomenon.

Feeling-type peoples' more emotional reactions to project events may baffle thinkers, along with their genuine concerns for other's

Table 10.5   Project Communications Issues Resulting from Thinking–Feeling Preferences

Thinking types may:	Feeling types may:
■ Prefer logic and justice as decision-making criteria	■ Prefer sentiment and caring as criteria for decision making
■ Appear cold and impersonal	■ Respond to others' emotional needs
■ Be unbothered by conflict	
■ Not provide emotional support when needed	■ Create a positive, supportive emotional tone on a project
■ Find it difficult to praise without criticizing	■ Be susceptible to emotional ups and downs
■ Create a negative project atmosphere	■ Inadvertently play favorites while attending to others' emotional needs

- ■ Many thinkers on a project can create a climate that is critical, unforgiving, and prone to conflict.
- ■ Thinkers may inadvertently hurt feelings of feeling types; the level of emotion of feeling types may baffle thinkers.
- ■ Thinking–feeling differences among people who work together often cause friction.

feelings and reactions. Feeling types working on a project can boost project morale because they make efforts to attend to other's issues and provide support. On the other hand, this concern for individuals can sometimes lead to favoritism: what's fair for the overall group is ignored while they care for the one person with a problem.

Perceiving–Judging

The perceiving–judging dimension of type is not so clearly connected with design and construction as the intuition–sensing dichotomy. Nonetheless, this dimension does exert some important influence on project communications and concerns (see Table 10.6).

This continuum describes peoples' preference for closure and moving on (judging) or their preference to keep their options open, continuing to explore new data (perceiving). Judgers tend to be more time-oriented and focused, preferring to:

Table 10.6 Project Communications Issues Resulting from Perceiving–Judging Preferences

Perceivers may:	Judgers may:
■ Avoid necessary scheduling and planning activities	■ Overschedule and overplan
■ Be able to break schedule constraints to do the job that must be done	■ Try to stay with schedules and plans even when those schedules and plans lose relevance
■ Be more flexible and open to change	■ Be less flexible and open to change
■ Be more spontaneous	■ Be more opinionated and judgmental
■ Value having fun	■ Work first, play later
■ Be able to get completely immersed in their work	■ Be more conscious of time than of the processes necessary to complete work
■ Be accused of being "wishy-washy"	■ Be accused of stubbornness

■ Many judgers on a project can create a climate that is critical, judgmental, and negative.

■ Judgers can often outargue perceivers, even when the judgers are wrong.

■ Perceivers frustrate judgers by arriving late and working in an unfocused manner. Judgers frustrate perceivers by trying to force them to follow a schedule and not give the work the detail it deserves.

■ Perceiving–judging differences among people who work together often cause friction.

■ Make plans and stay with them

■ Make and follow schedules

■ Use and synchronize watches and clocks

■ Get to places and events early

From their preference for closure, judgers also usually are more opinionated than perceivers. While people may accuse judgers of being overly opinionated, judgmental, critical, and argumentative, they accuse perceivers of not being judging enough, i.e., being "wishy-washy."

Perceivers tend to prefer to:

- Keep their options open, avoiding rigid plans
- Take priorities one at a time, following the issues rather than a set schedule
- Ignore clocks and watches because they represent an artificial, forced sense of priorities

Judgers frustrate perceivers by trying to get them to follow plans and schedules that the perceivers think are useless. Perceivers frustrate judgers by arriving late and by missing time commitments.

On a design and construction project, judgers will help force the project to be on time but they may do so at significant costs of quality and productivity. Perceivers may take longer but they are more likely to seek out and resolve quality problems.

Temperaments

The four different dimensions of preferences result in 16 different possible types [e.g., extrovert(E)–sensing(S)–thinking(T)–judging(J), or ESTJ; introvert(I)–intuitive(N)–thinking(T)–perceiving(P), or INTP], each of which has its own profile of specific preferences, attitudes, beliefs, priorities, and communications styles. While it is not possible in this text to provide details on all of the sixteen type combinations, it is worthwhile briefly exploring four of the chief combinations of preferences, or temperaments. The temperaments explain a great deal about recurring project conflicts and provide insight useful for conflict resolution (see Table 10.7).

The four major combinations of temperaments are as follows:

- Intuitive–thinking, including INTJ, ENTJ, INTP, and ENTP
- Intuitive–feeling, including INFJ, ENFJ, ENFP, and INFP
- Sensing–judging, including ESTJ, ISTJ, ESFJ, and ISFJ
- Sensing–perceiving, including ESTP, ISTP, ESFP, and ISFP

Intuitive Thinkers (NTs)

NTs comprise about 10 percent of the U.S. population, but many architects are NT. NT combines intuition with thinking, resulting in systems thinking and visual thinking.

Table 10.7 How Workers with Different Temperaments Tend to Communicate on Design and Construction Projects

Intuitive–thinking (NT)	Intuitive–feeling (NF)	Sensing–judging (SJ)	Sensing–perceiving (SP)
▪ Typically work as designers, planners, or owners of a subcontracting firm working with new technology	▪ Typically work as architects or as general contractors, often in corporate real estate as facilitators or entrepreneurs	▪ Typically work as contractors, site superintendents, subcontractors, or engineers	▪ Typically work as craftspeople, tradespeople, or detail engineers
▪ Visualize the key concepts	▪ Create a mood, often positive	▪ Organize details	▪ Get immersed in craft
▪ Make strong design statements	▪ Respond to others' emotional needs	▪ Spot design flaws and problems	▪ Avoid paperwork and routine
▪ Project arrogance (though they may feel self-critical)	▪ Struggle with detail and documentation	▪ Want more detail from the intuitive types	▪ Upset SJs by not sticking to schedules
▪ Can intuitively come up with systems solutions to problems	▪ Build relationships with clients	▪ Question the intuitives' intelligence	▪ Are subversive in a quiet way
▪ Try new technologies	▪ Build the project team	▪ Make rules for the other temperaments to follow	▪ Are spontaneous
▪ Are intrigued by the novelty of the idea of partnering but skeptical of interesting results	▪ Can design spaces that create a feeling	▪ Are worried about how partnering will impact established procedures and hopeful it will increase adherence to rules	▪ Are energized by emergencies
	▪ Are most interested in and optimistic about partnering		▪ Are willing to give partnering a shot, but uncommitted to results one way or the other

The architect in *The Fountainhead* is portrayed as an intuitive thinker. The popular television character Frasier (and his brother Niles) are both drawn as NTs. Like most NTs they are most concerned with demonstrating their own competence, not to show off but just for their own satisfaction.

NTs lead their design firms to work with new technologies and to experiment with new approaches to design. They favor strong, clear design statements. In contracting and engineering, they try to offer comprehensive services and long-term planning.

NTs tend to:

- Value intelligence, their own and others'
- Be interested in concept, ideas, and abstraction
- Enjoy, read, and collect (but not always finish) books
- Be perceived as (and may be) arrogant, yet also be highly self-critical
- Be able to visualize large-scale systems solutions to problems
- Think in pictures
- Favor obscure humor, science fiction, and computer games
- Be interested in the concept of partnering, but skeptical of any truly interesting results

Intuitive Feelers (NFs)

NFs also comprise about 10 percent of the U.S. population. Primarily driven by an interest in relationships, they intuit the feeling side of decision making. NFs are interested in and usually know how others feel.

Many NFs find their way to architecture and design, and bring with them a great sensitivity to "process." We have also encountered a number of NF general contractors, people who like to (and are able to) build a strong project team and get involved with the whole process of construction. Their firms are likely to be highly participative and they can easily develop participative ways to work with clients. Client involvement is usually a priority for NFs. Detailed work, infrastructure, and record keeping, unfortunately, are not NF strengths.

NFs tend to:

- Want to understand and respond to clients' needs
- Build strong relationships with peers and clients
- Be able to create designs that create a feeling, e.g., restaurants in which one feels comfortable, hungry, etc.
- Struggle with record keeping and documentation
- Idealize relationships
- Seek meaning in many activities
- Be most interested in the practice and follow through of partnering, and most optimistic about its results

Sensing Judgers (SJs)

SJs represent about 40 percent of the U.S. population. Primarily focused on responsibilities and what one *should* do, they combine an eye for detail with a judging preference. They have and may express strong opinions on details and procedures.

SJs live in a world of rules and procedures they have largely created for themselves. They establish procedures for eating breakfast, parking their car, and organizing their closets. They are naturally organized and often uncomfortable when details are chaotic.

SJs bring order and organization to the potentially chaotic world of design and construction. An SJ principal elevated to office manager in a large engineering firm led by intuitive people began the first week in her position by issuing work rules, standard hours, and an organization chart. On the job, the site superintendent is often an SJ as are many contractors, engineers, and subcontractors, all people with an eye for and an appreciation of detail.

SJs tend to:

- Understand and work comfortably with details, procedures, and rules
- Implement and finish what they start
- Easily make order out of chaos
- Work on things one task at a time

- Be uncomfortable with what they can't see
- Sometimes ask for more detail and documentation than is necessary
- Work first, play later
- Use lists extensively to get organized
- Favor the humor of one-liners (ISTJs seem to favor black humor)
- Read contracts literally and be unforgiving of errors

Sensing Perceivers (SPs)

Sensing perceivers begin with the same eye for details as SJ. However, instead of linking that with the closure of the judging function, they link it with the open-mindedness of perceiving. The result is a temperament that is curious, nonjudgmental, spontaneous, and somewhat unpredictable.

SPs have the ability to get completely immersed in whatever they're working on and to practice and rehearse until it is perfect. The Nike "Just do it" commercial could be the SP anthem.

SPs are not much interested in procedures or rules, though. Their crafts, perfection, and/or immersion orientation to their activities runs at cross purposes to schedules, rules, and procedures. SPs want to do it their way, and their way is unpredictable.

SPs predictably avoid clerical or administrative work in large corporations but gather in large proportions in the buildings trades. It is the trademark of an SP carpenter, electrician, or plumber to arrive late, take a long lunch, and then still complete the project in an absolutely artistic way in the eleventh hour (while the organized, scheduled SJ supervisor is reeling with anxiety).

SPs tend to:

- Get immersed in what they do
- Focus more on quality than on schedules
- Feel boxed in by plans
- Change their minds easily as they take in new data (which is hard for them to stop)
- Have a difficult time adhering to rigid schedules and rules
- Value having fun first, then getting down to business

- Be at their best in emergencies, and sometimes help create emergencies by waiting to get to work until the last possible moment

Bibliography

Well over 100 books are available providing more detail on the Myers–Briggs Type Indicator. Books are available in all the topic areas in which people use the MBTI: management training, team building, career development, marriage counseling, and religious issues. Book lists, MBTI training information, and the books themselves are available from two organizations: Center For Applications Of Psychological Type (CAPT), 1–800–777-CAPT, or Type Resource Inc. (TRI), 1–800–456–6284. The books our clients find most useful are the following.

William Bridges, *Organizational Character* (Consulting Psychologists Press, New York, 1993). Bridges applies the MBTI to develop character profiles of whole organizations. Useful reading when trying to understand client organizations or the culture of particular project teams.

Alan Brownsword, *It Takes All Types* (San Francisco, 1991). A readable, yet deep and thoughtful explanation of types and temperaments.

David Kiersey and Marilyn Bates, *Please Understand Me* (Prometheus Nemesis Press, San Francisco, 1984). The best-selling classic book on type, with outstanding descriptions and analysis of the temperaments. A bit dated these days, however.

Lavon Neff, *One Of A Kind*. Neff describes how type plays out in children and childrearing. Of interest to parents but also of great general interest because of the clear descriptions of the enduring characteristics of the types.

11
Training
Improving One-on-One Communications Skills

Summary

The word *partnering* conjures up images of people sitting around a table, intently analyzing project problems and devising creative, strategic solutions. That image is partially accurate.

For many, the core of partnering is more in the everyday communications project team members have with one another on the job site, in their offices, and over the phone. Everyday one-on-one communications impact partnering in two ways:

- They are the arenas for implementing much of what is written in partnering documents (*goals statements, communications procedures, conflict resolution,* etc.). If the one-on-one communications are ineffective, it will be difficult to implement the documents.

- They embody and reflect the spirit and intentions of partnering. It is not possible to have partnering with great documents but with a project team that treats its members with a high level of antagonism and disrespect.

This chapter:

- Describes in more detail the role and impact one-on-one communications have on overall project communications

- Describes a collaborative approach to one-on-one communications that reflects partnering goals
- Provides a specific example of how this collaborative approach resolves typical project miscommunications

Perspective on Project One-on-One Communications

Everyday project communications are both an opportunity and a problem. They are an opportunity in that the way people act in them is completely under their own control. No external influences dictate how people communicate. They can make their own rules and create the overall project culture they want.

The problem is that many people in design and construction do not have strong skills in a collaborative style of communications. They are more comfortable and familiar with aggressive rather than assertive communications, with a win-at-all-costs rather than a win–win approach.

Design and construction practitioners are beginning only recently to work seriously with the kinds of collaborative skills their corporate clients have used for years. Most large corporations have internal training departments that train managers and employees in more collaborative styles. Design and construction practitioners are coming to learn these styles for the same reason the corporations have: they are essential tools for organizational productivity.

One-on-one conversations play a significant and difficult role in design and construction projects:

1. *Disproportionately influential.* Many project conversations take just a few minutes but exert a great deal of influence over the next day's and even the next week's work. People typically talk with each other when:
   - They have finished one phase of their work and need approval, review, monitoring, sign off, and guidance for the next step. A carpenter may work on a project for several days, then have her supervisor review it. In one or two minutes the supervisor will communicate if the work is acceptable and what to do next.

- They see something going wrong in someone else's work and they feel it necessary to speak out.
- They offer guidance, help, and encouragement.
- They ask for guidance, help, support, and advice.

2. *Emotional.* Most of the situations listed above are emotional. People have a great deal riding on these situations, so they may not react positively. Even people who genuinely invite criticism may have a hard time accepting it.

3. *Elusive, intangible.* Communications are difficult to quantify or count, especially for people more used to working in the tangible world. When something goes wrong in communications, people don't usually recall the pattern of information exchange. Instead they remember the more general (and less helpful) sense that the other person was unreasonable or argumentative.

4. *Organizationally bounded.* The lines of trades, professions, and organizational hierarchies are clearly drawn in design and construction. Any conversation that crosses those lines carries the weight of both peoples' professional biases and prejudices. When an architect and a contractor argue over details in a drawing, it's not just the two of them but decades of professional prejudice confronting each other. The abstract, arrogant, sloppy, spacy architect (as seen by the contractor) goes up against the stubborn, rigid, small-minded, obsessive contractor (as seen by the architect).

5. *History of win–lose conflict.* All trades and professions have their myths and legends. Part of the lore of the design and construction business is the "tough nut" characters, whether architects, engineers, or contractors, the people who "stuck to their guns," "knew their own mind," "wouldn't give in." Some of the most revered designers, engineers, and contractors in the country today are known as much for their stubbornness as for their technical and professional skills.

All this is to say that there are not many role models in real estate, design, engineering, and construction for a collaborative style of communications. Individuals may be decent, intelligent, skilled, and thoughtful. Yet the established professional role models often reflect a more inflexible win–lose stance and a style that rewards rather than resolves conflict.

General Perspective on One-on-One Communications

Outside design and construction, more general problems plague one-on-one communications (Table 11.1):

Fight-or-Flight Impulses. Faced with awkward or difficult communications, many people respond in an "either/or," fight-or-flight way. They either attack, take over, or try to control, or they fold or walk away. The whole middle ground of reasonable, objective, creative discussion gets lost in the heat of argument.

Table 11.1 Perspective on One-on-One Communications

On projects one-on-one communications are often difficult because they are:	In general, forces that make one-on-one communications difficult include:
■ *Disproportionately influential:* A ten-minute conversation can decide the course of work for several days or even weeks ■ *Emotional:* With so much riding on conversations, they can easily become emotional ■ *Elusive:* It is difficult to recall exactly what happened and to get data on conversations ■ *Organizationally linked:* When two people on a project talk, it is really firms, trades, or organizations talking ■ *Burdened by a history of conflict:* The history of design and construction is rooted in conflict with few models of one-on-one collaboration	■ *Fight-or-flight impulses:* In handling differences of opinion, many people respond with an either/or, fight-or-flight impulse, skipping the middle ground of reasonable discussion ■ *Urge to control:* Many people react to the fact that they cannot control the outcome of a conversation by trying to force control ■ *Unconscious patterns:* Many conversations, particularly recurring ones, often fit patterns of alternating topics, timing, attitude, and mood shifts ■ *Projection and transference:* People often project onto others (especially authority figures) patterns of conversation they are familiar with since childhood

Urge to Control.   People in a conversation, especially any of the project examples listed above, are not fully able to control the outcome of the conversation. The outcome depends in large part on the other person. Being unable to exert control is always awkward and uncomfortable. If the issue being discussed is an important one, then the feelings of awkwardness and discomfort may be extreme. People understandably respond by trying to control, and that may work in technical problems. In conversations, however, trying to control another person inevitably leads to problems and cannot work.

Unconscious Patterns.   When people meet on a recurring basis their conversations often fit into recurring patterns. The patterns include sequences of topics, amount of time spent on a topic, attitude, and mood shifts. The relevance of this point for conversations on a design and construction project is that it explains why conversations sometimes seem to have a life of their own. The more one becomes aware of these patterns, the easier it is to manage them.

Projection and Transference.   When one-on-one conversations go awry there may seem to be no logical explanation. On the surface, both people seem like rational adults. The problem is below the surface, in that one person may be projecting onto the other patterns of conversation familiar from one's own experiences.

This occurs most often when one is communicating with an authority figure, and one projects onto that person patterns of communication from interacting with one's parents. This occurs in design and construction process situations where one person has the ability to get under someone else's skin. Often the person subconsciously reminds one, with a tone or a gesture, of the interactions one had with a parent.

Lack of Training and Development.   Despite extensive available research and data on effective communications, relatively few people participate in communications training. Established academic training gives much more weight to technical and professional training than to communications training. Indeed it is quite possible to get through secondary school and college with no formal training in communications, even though communications skills are one of the most important contributors to overall job success.

Attitudes and Actions

The work of strategic thinking is complete, the procedures for identifying problems and resolving them are clarified, a set of steps is agreed upon to escalate conflict in a thoughtful way, and empowerment principles are in place. The next phase of conflict resolution makes it possible for all the previous work to be translated into everyday life on the project.

Without a clear personal system of workable assumptions and an effective set of collaborative communications behavior and skills, people on the project team will have a hard time living up to and implementing even the best conflict resolution procedures. Without these assumptions and skills, they will also be unable to resolve the new conflicts that emerge between the cracks of the thoughtful procedures and processes.

Yet there is not much understanding of what a collaborative approach to conflict resolution really is. Some people think it still means getting your own way but with a smile instead of a shout. Others think it means giving things up and learning to live with compromise. A truly collaborative approach to conflict resolution need not be any of these—it could be much more interesting and productive.

In the following sections we describe the assumptions people hold that lead them to ineffective work in resolving conflicts, and we describe some of the characteristic phrases typically heard in unproductive conflict discussions. We then describe more productive, collaborative assumptions and characteristics (see Table 11.2).

Effective individual action in resolving conflicts is much more than a set of assumptions and catch phrases, however. In order to function optimally in resolving conflicts, it is usually necessary to develop some different communications skills. Many people, and especially people who work in design and construction, find these skills potentially very useful but difficult to acquire without some practice.

Unproductive Assumptions

When people in any situation disagree their best thinking, creative imagination, and logical reasoning powers short-circuit while an almost biological fight-or-flight impulse takes over. Faced with a conflict, even the most logical of people find their thinking polarized around win–lose assumptions, their behavior gravitating to

Table 11.2   From Standoff to Collaboration in Conflict Resolution

Unproductive assumptions	Tip-off unproductive phrases
■ Failure	■ "I know how you feel."
■ Problem	■ "I agree" or "I disagree."
■ Pointless	■ "I've said my piece."
■ Fight	■ "Trust me."
■ Right answer	■ "Calm down."

Create a Structure Conducive to Resolving Conflicts

- *Defined time:* Schedule a series of one- to two-hour talks to work on the problem.
- *Ground rules:* Neither person can win at the expense of the other person or the project.
- *Problem–solution focus:* Focus on the problem and solution, not on blame, revenge, or ego.

Productive assumptions	Tip-off productive phrases
■ *Inevitable:* Conflict is inevitable, not an indication that someone has failed.	■ "Tell me more about how you feel."
■ *Opportunity:* Resolving conflicts provides real opportunities to devise more effective solutions for both people and for the project.	■ "Forget agreement or disagreement. Let's look at the data and explore alternatives."
■ *Problem solving:* Conflict resolution is not pointless or a fight, but solving problems.	■ "I've said my piece but that's only half of it. What do you think?"
■ Exploration orientation, not "right answer": Working together on the problem will lead to more lasting solutions.	■ "Don't trust me or not trust me. Let's look at the data and examine the problem."
	■ "You seem pretty upset. Tell me more."

attempts that are neither very sophisticated nor very effective to control others and get what they want.

At this point, people hold a number of assumptions that can often be heard in discussions on the job site.

The "Failure" Assumption.   "We have conflict; we must have failed." Although people logically admit that conflict is inevitable, there is a part of many people that holds the assumption that conflict is an indication of personal failure.

This assumption can underlie two different behaviors that can make conflict resolution more difficult. First, this assumption can lead back to the person holding it, where it turns to guilt, shame, and a shaky stance and low-quality thinking during discussions of conflict.

Alternatively, the failure assumption can also lead to a blame orientation. "If there's a failure here and I didn't cause it, then it must be your fault!" This orientation turns what could be effective conflict resolution into a witch hunt.

The "Problem" Assumption.   "Oh, boy, we've got a conflict now. This is going to be a real problem." People who have little experience with effective conflict resolution (except possibly negative experience) have an understandably difficult time seeing the positive side of conflict and the perspective that conflict is inevitable, that it is not necessarily anyone's fault, or that it may often be an opportunity to develop more effective solutions to a shared problem.

Nonetheless, the problem assumption leads to self-limiting thinking and behavior in conflict resolution discussions. It is difficult to give a conflict one's fullest attention when one doesn't really think the effort is going to go anywhere anyway.

The "Pointless" Assumption.   "Why are we going to discuss this? Nobody ever really worked out any solutions to real problems in discussions. Nothing is going to come of this."

People who have limited successful experience with conflict resolution may justifiably assume that, in fact, conversation is pointless. After all, they have no experience to lead them to think otherwise. We find this assumption to be surprisingly prevalent among people in the design and construction fields where the work is highly segmented and where it is possible to get a measure of the work done without really having to interact with the workers. Nonetheless if people carry this assumption into conflict resolution discussions, they will limit their ability to explore alternatives to define new approaches to solving the problem.

The "Fight" Assumption.   "If I'm going to get what I want, I'm going to have to fight for it, and possibly fight hard." This is another understandable assumption in a business that for decades favored and rewarded people who could argue well on their own behalf, whatever the consequences may have been for others.

In conflict resolution, this assumption leads to a combative stance that can easily derail a conflict resolution discussion. It is difficult to work constructively on a problem with someone who is always fighting hard for their own way.

The "Right Answer" Assumption.   "I've got the answer." In most cases it is useful to begin discussions with some kind of an action step or result in mind, but in conflict resolution this can be destructive. When an individual begins with an "answer" in mind, it can make it more difficult to pay real attention to others' ideas, or even to examine the problem in any depth. Why bother exploring the problem if one already has the answer?

Besides, most problems that turn into conflicts have many "answers." When people advocate for their own answer, it blocks exploration of alternative and perhaps superior approaches.

Tip-Offs of an Unproductive Discussion

Listening to a conflict resolution discussion from the outside, it is often possible to detect indicators of the discussion's effectiveness. A handful of phrases, unfortunately common phrases, indicate with some accuracy that a discussion is not very productive:

- *"I understand you. I know how you feel."* People often make this comment to each other to show empathy and understanding for the other person's position. Unfortunately, the expression frequently has the opposite impact. Almost inevitably, when one person claims understanding of the other, the other thinks, "Oh no you don't!"

- *"I agree" or "I disagree."* In conflict resolution, agreement or dis-agreement is not terribly relevant. People attempting to resolve a conflict often feel compelled to express their agreement or dis-agreement with the other person at every step of the discussion, perhaps out of a fear of losing ground, perhaps in the interest of expressing solidarity with the other person. Either way, the more a discussion focuses on agreement or disagreement, the less it focuses on problem solving. When two people are more oriented to examining the problem and devising creative solu-tions to it, the whole subject of their agreement or disagreement becomes less and less a factor.

- *"I've said my piece."* People often say this at the conclusion of an ineffective discussion. They usually say it defensively and self-righteously, as in "I did all the right things; it is the other person who is to blame for the problem."

  The "I've said my piece" approach indicates, however, that the person thinks the conversation is about presenting an opinion for approval or disapproval rather than participating actively in a two-way dialog. Saying one's piece is only half of the discussion. After one says it, what does the other person say? Where does the discussion go from there? What comes next?

- *"Trust me."* This is an expression one associates more with used car purchases than with design and construction work, but one hears it often when people who have a difference of opinion try to get each other to understand their own point of view. It is an understandable request but usually doomed to fail. If one has to ask for trust, it's probably too late to really do anything about the problem.

- *"Calm down."* This is the grandparent of conflict resolution short-circuits. As with the other phrases, people often express this one with the best of intentions but seldom witness it to pro-duce any positive results. It usually backfires with a reply of "Don't tell me to calm anything!"

The Miracle of Conflict Resolution

Time after time, against seemingly impossible odds, conflict reso-lution in partnering works. The odds against resolving real conflicts on a project are staggering. People have real points to make, real arguments to explain, and real money to lose. They may have spent days, sometimes weeks, even months trying to resolve a conflict on their own without any success. The situation has not improved at all; if anything, it may well have gotten worse as time dragged on.

The people involved in the conflict have tried to be flexible, understanding, and reasonable. Still, there is no resolution to show for their efforts, no solution to the problem.

What, then, does it take to make conflict resolution effective? It helps to begin with some structure:

- *Defined time.* Blocking out a finite amount of time to really work on a resolution.

- *Ground rules.* Following, for that time period, the following rule: "Neither person can win at the other person's expense, and both people can't win at the expense of a mediocre solution." In other words, for at least a finite period of time, neither person can simply reiterate his or her own position without yielding.

- *Problem–solution orientation.* Focus on the problem, not on each other's positions.

Productive Assumptions

Within this groundwork, more productive assumptions about conflict and conflict resolution include:

Inevitability, Not Failure.   Conflict in design and construction does not necessarily mean that people have failed. Instead, conflict is unavoidable and inevitable. Too many people depend on too many other people in an unpredictable environment to have conflict-free work.

Opportunity, Not a Problem. The only real problem involved in conflict is that it takes some time to work out. The time taken to work it out, however, can just as well lead to better solutions and new approaches, not just a compromise. Effective problem solving often involves thinking the problem through at a level of depth that people did not get to in their initial planning of the project.

Problem-Solving Orientation, Not Fighting or Pointless Attitudes. There is a real problem to solve here, so it is not pointless at all to work on it. Working on that problem, I devote my efforts in conflict resolution towards finding the solution, not towards fighting to defend my own point of view.

Exploration Orientation, Not "I Have the Right Answer." If I begin with "the right answer," I make it difficult for others to contribute and own the results. Thus I channel my participation into examining the problem along with the others.

Another useful way to work through the solution dilemma is to work with solutions other people suggest as if they were one's own. Instead of trying to get the other person to explain or defend their ideas, this involves accepting the other person's ideas and exploring them actively and with a fully open mind.

Tip-Offs of a Productive Discussion

Just as a number of phrases indicate that conflict resolution discussion is probably not productive, the following phrases indicate that it probably is productive:

- *"Tell me more about your feelings."* People in design and construction have a difficult time discussing feelings; emotions don't seem relevant or professional. Yet a period of time spent listening to each other's feelings without making any judgments can help two people get to the underlying reasons behind the feelings quickly. Listening to feelings also helps take the tension and pressure off a conversation, building trust.

- *"Let's explore the reasons behind our disagreements."* Discussing and exploring reasons rather than opinions will lead more quickly to solving problems.

- *"I've said my piece but that is only half of it. What do you think?"* Instead of stressing only what one has to say, it is helpful to work equally hard at trying to understand what the other person has to say. Many people think they are accomplishing something in a conversation only when they are talking. Much of the real work in thinking in a conversation, however, takes place as a result of listening.

- *"You don't have to trust me or not trust me. Instead, let's examine the data and explore possible new actions."* It is not so difficult to intend to focus on the problem instead of on the people but much more difficult to bring that intention to everyday conversations. This phrase indicates that people are moving in the right direction.

- *"You seem upset. Tell me more."* Instead of telling people to "calm down," this approach invites the person to work through the upset underlying the problem and work towards the reasoning and the sources underlying the problem.

How Partnering Communications Training Works

The perspectives on one-on-one communications provide the foundations for an approach to communications skills training for partnering in a three-step process (see Table 11.3):

Step 1. *Clarify and anticipate difficult specific situations.* Communications are elusive, but it is quite possible to predict and antici-

Table 11.3 Communications Skills Training for Partnering: A Three-Step Process

Step 1. *Clarify and anticipate difficult project situations.*
- It is possible to predict problems.
- By predicting problems, it is more possible to handle them effectively.

Step 2. *Identify and break recurring problem conversation patterns.*
- When conversations don't work they often fit recurring patterns.
- Specific conversation skills can break these patterns.

Step 3. *Model collaborative resolutions to problems.*
- Since people are more familiar with win–lose approaches, it is helpful to model and play out specific examples of collaborative resolutions to problems.

pate the problem conversations that will evolve on a project. Step 1 of improving communications is to list those problem situations.

Step 2. *Identify and break destructive patterns of communications.* Since problem conversations fit patterns, it is possible to identify in advance what some of the most destructive patterns of communications on a project will be. Once these patterns are identified, it is possible to break and manage them.

Step 3. *Model successful resolutions in advance.* Since people are used to win–lose patterns, it is helpful to sketch out specific examples of what successful collaborative resolution to problems would be.

Step 1. Clarifying and Anticipating Difficult Project Situations

Because one of the major problems in working with one-on-one communications is their elusiveness and intangibility, it is useful to start working with communications skills by clarifying and anticipating awkward situations.

On design and construction projects it is often quite easy to quickly list current and future awkward communications. Many of the same issues come up on each job. Also, just by looking at a drawing, many design and construction practitioners can predict who will argue with whom about what in the coming days.

Typical predictable problem conversations include:

- Contractor complains to architect about mistakes in drawing
- Architect complains to contractor of excessive nitpicking and overreaction to mistakes
- Anyone gives bad news to the client (there is a delay, someone made a mistake, the material is no longer available, etc.)
- Anyone gives anyone constructive criticism
- One person asks a favor of another person
- One person invokes a rule or procedure the other thinks is unimportant or nonexistent
- Person A complains to person B about person C

- Anyone gives bad news to anyone
- Anyone acts as a messenger
- Anyone admits a mistake
- Anyone defends a course of action one thought was correct

Simply identifying predictable problem areas has a sobering effect on a project team, much the same as when a driver knows the location of the potholes on a highway. The potholes don't go away, but the driver approaches them more carefully.

Step 2. Identifying and Breaking Problem Patterns

When conversations fail or fall short, they often do so in patterns that are predictable and easy to observe. After clarifying the difficult situations, a useful next step is to identify the dysfunctional patterns that can arise. Once one can see the patterns, one can break them and manage them.

To identify the patterns that occur in troublesome conversations, it is usually necessary to work both with people who are directly involved and with several outsiders who may have a more objective perspective on what is occurring.

The project team works in small subgroups, each one taking a few of the problematic conversations and exploring possible patterns that may arise. Typical examples that they find on a project are:

- Architect and contractor are locked in a "mutual blame" pattern in many of their conversations. The contractor blames the architect for sloppy work, recurring mistakes in detail, and lack of respect. The architect blames the contractor for unwillingness to read anything into the drawings, for asking for too much detail, for reacting inflexibly to the architect's on-the-spot ideas and solutions.

- Architect, contractor, and client are locked in an "elusive" pattern for making decisions. The client sees things being constructed, doesn't like the way they look, requests changes, but then acts surprised about having to pay for them.

- Contractor and subcontractors are locked in a "Who, me?" pattern of taking responsibility. As changes evolve in the project, the subcontractors request exorbitant add-on fees to cover their time and materials. When accused of exploitation of the changes, they reply with the classic "Who, me?"

- Subcontractors are locked in a "I gave up last time, now it's my turn" approach to resolving rescheduling priorities. They cite small "gifts" they have given and losses they have taken earlier on the project as reasons for current inflexibility on a major issue.

Step 3. Modeling Collaborative Resolutions

One of the most effective ways to break the destructive patterns once they have been identified is to replay them with the pattern broken. To do this, we return to the same three-part format used in the conflict resolution and group communications chapters of this text (see Table 11.4).

1. Define Aspects of the Problem. Many times people fall into patterns of unproductive communications and lose sight of the problem they are trying to solve in the first place. In particular, people often describe the problem in general or universal terms: "We have a communications problem," "The drawings always have mistakes," or "The contractor is inflexible." We ask people first to list the numerous aspects of the problem that make it a problem, moving from general and universal to much more specific descriptions.

This first step:

- Breaks the pattern.
- Refocuses on the problem.
- Breaks the "either/or" feeling by focusing people on the multiple aspects of the problem.
- Places people in an inquiring rather than an enforcing frame of mind.
- Places people on the same side of the table.
- Moves the problem-solving process from general to specific, which is essential. It is much easier to solve specific problems than it is to solve general ones.

Table 11.4   Collaborative Problem Solving: A Three-Step Approach

1. List aspects of the problem	2. List possible action steps	3. Choose several specific things to try
■ Take time to define the problem ■ Move from general descriptions of the problem to much more specific ones ■ Find a workable solution by listing all aspects of communications problems ■ Don't rush to a solution, it will probably be effective only in the short term ■ Realize that both sides contribute to the list	■ Brainstorm, which means list anything, you can get selective later ■ Realize that there are no bad ideas in this phase ■ Focus on the problem, not on blame ■ Push yourselves to see if you can come up with new ideas ■ Focus on ideas that don't make either person a loser ■ Use the problem aspects list	■ Choose several possible steps ■ Define specifics: who will do what, when, where, and how ■ Sign what you agree to ■ Plan to meet again in a short period of time to monitor and refine you plans

2. Brainstorm Possible Actions. Here again we focus on possibilities to take the pressure off peoples' either/or tendencies. Brainstorming possible actions:

■ Keeps the destructive pattern broken.

■ Focuses on the problem, not blame.

■ Pushes both sides into thinking rather than defending.

■ Makes it clear that many different actions may address the problem.

3. Try Several Action Steps. Instead of ending the conversation with a judgment or a win–lose proposition, we try to end with an experiment, some things to try. The things to try are not mere possibilities or wishes; they are definite, mutually agreed upon specific action plans with time, location, and results mapped out by and for both sides. This approach:

- Focuses on improvements rather than blame.
- Communicates the intention of continuous improvement.
- Eliminates winners and losers.
- Reinforces multiple solutions to problems.
- Implies follow through to review, refine, and monitor.

Examples of Collaboration

We provide numerous detailed examples of collaborative problem solving in the Case Studies section of this book in Chap. 15.

Can People Change? Do People Learn and Use One-on-One Communications Skills?

When we provide one-on-one communications skills training as part of a partnering project, people in the group usually ask these two questions. We provide communications skills training to thousands of managers, professionals, and tradespeople every year. We have the opportunity to work with many of those people several times. We have provided management development in some companies for groups of managers, several days a year, for six and seven years. Thus we do have some observations and opinions in answer to the two questions that head this section.

1. *People don't change who they are, but many people develop and apply new communications skills.* Why program participants ask the general question "Do people change?" is unclear; perhaps it is because communications skills so closely define a person's character and personality. Working with people over time, it does not seem that many (or any) change the core of their characters or personalities.

However, many people we work with do learn new communications skills, and they learn to apply them very effectively. We chart participants' progress in many companies with internal, confidential perception surveys completed by participants' subordinates, peers, superiors, and customers. We note in the survey data and in our own direct work with people that many do add new skills.

2. *Partnering does not ask people to change their personality or to develop a high level of new communications skills, but to eliminate the rough spots and acquire some new techniques.* We would enjoy training partnering workshop participants extensively in collaborative communications, and they would likely find it interesting and worthwhile. However, partnering does not attempt to provide such extensive training, nor is such extensive training necessary for partnering to be effective.

The amount of one-on-one communications skills training necessary for partnering to be effective focuses primarily on what is not acceptable. Partnering participants will not and need not become highly skilled at collaborative problem solving, but it is necessary for them to know:

- That aggressive one-on-one communications styles undermine partnering.
- That there are other ways to solve problems aside from shouting more loudly.
- That here are some basic techniques, as outlined in this chapter, they can draw on to make sure their everyday one-on-one communications support partnering goals.

Bibliography

Most books on communications skills describe very similar approaches to collaborative communications. Specific titles program participants tell us are useful to them include:

Stephen R. Covey, *The Seven Habits of Highly Effective People* (Fireside, New York, 1989). Covey provides not skills but a valuable perspective on why a collaborative approach is more effective.

Roger Fisher and William Ury, *Getting To Yes* (Penguin, New York, 1985). Widely hailed as the classic guidebook for win–win communications.

Roger Fisher and Scott Brown, *Getting Together* (Penguin, New York, 1988). Detailed, more specific sequel to *Getting To Yes*.

Harriet Lerner, *The Dance of Anger* (Harper & Row, New York, 1985). Billed as a "women's issues book," this book in fact provides excellent insights into destructive communications patterns that arise on every construction job site and skills useful in developing more productive working relationships.

12
Better Meetings

Summary: Meetings

Most partnering *goals statements* establish a tone of cooperation and trust and spell out project team members' hopes to communicate effectively over the course of the project. One specific area they can focus on to improve their overall communications is in the various kinds of meetings they have.

On many projects, three different kinds of meetings, weekly, daily, and informal, can significantly improve overall project communications. If all three of these meetings are brief and well-organized, they can save many hours of individual work, increase communications efficiency, and avert conflicts. Because effective meetings can play an important role in improving overall project communications and information flow, we consider them to be a major part of the infrastructure of project communications (see Fig. 12.1).

Weekly Project Meetings

Most projects hold weekly meetings of key project team members and decision makers, but the meetings are not always run well. Often, people attend but do not get to discuss the chief issues that concern them. People who are not comfortable participating may get left out even when they have valuable information. People who like to talk may dominate. Meetings lose focus and take longer than necessary.

1. Weekly project meetings
   - Define, clarify meeting's purposes.
   - Determine who should attend.
   - Determine when and where meetings should occur.
   - Develop a template agenda to define recurring issues for discussion.
   - Develop ground rules for effective discussion.
   - Define leadership tasks; involve members in leadership tasks.
2. Daily "check-in" meetings
   - Define purpose and content.
   - Specify location, time, and attendees.
   - Use flip chart to record action items.
3. Informal meetings
   - Identify pairs and trios of project team members who have predictable needs to exchange information.
   - Establish regular schedule of brief, frequent meetings.

Figure 12.1.   Procedures to develop for meetings.

Working on *communications procedures,* the project team members draw on their collective experience and examine the communications needs of the project to spell out what it will take to make the weekly project meetings productive. Usually their considerations include the following:

- *Why are we meeting?* Devising effective communications procedures must begin with clarifying the purposes of a weekly meeting. The more clearly the purposes can be defined, the easier it is to design procedures to attain them. Usually weekly meetings have several purposes, including:

  Updating all project team members about key events, situations, problems, and opportunities that are occurring for other team members and that may impact their own role and tasks.

  Contributing information from multiple trades, professions, and disciplines about problems and opportunities that the project faces.

Making decisions to enable the project to proceed to next phases of work.

Solving problems that cross trade or professional lines.

In specifying purposes such as these it is useful to consider the types of information that people need to give and receive at different phases of the project. If the purposes of the meeting are clearly defined, the meeting can efficiently manage the timely exchange of crucial information among the project team members.

- *When will the meeting occur?* Often one particular time and day a week is preferable because it enables people both to collect needed information and make key decisions enabling the project to go forward.

- *Where will the meeting be held?* Meetings during construction are usually held in the trailer, but at times people find it is useful to precede a meeting by doing a brief walk through of the job site. This provides first-hand contact with many of the issues people will be discussing later.

  On the other hand, people may find it useful at times to meet in the offices of one of the firms working on the project or in the client's office. It may be useful to meet in the client's offices, for example, if several people in the client organization need to have their input included in the project discussions at a particular time.

- *Who will attend?* "The last segment of the work is completed; the project is ready to move ahead. We're just waiting for approvals from 'Person X.'" Unfortunately, no one seems to be able to get that person to a meeting.

  Finding the right people to provide input and approvals at the right time can make or break a project. People at a partnering workshop attempt to anticipate when key people need to be involved and to schedule them well ahead of time.

- *What gets discussed?* If members of the project team think about it a bit ahead of time, they can usually spell out what items will need to be discussed at most project meetings. The items on this list of recurring issues constitute the content of a recurring or *template agenda.*

  If participants think about the items on the list in more detail, they can also often agree that meetings will be most productive if the project team addresses items in a particular sequence. This sequence provides the order for the template agenda.

The template agenda will not cover every item for every meeting. Actually, the person leading the meeting will usually make a few refinements to it for every meeting. However, agreeing on a recurring meeting form helps ensure that meetings address key issues while also providing some continuity from one meeting to the next. Participants often find it easier to contribute when they know how a meeting will evolve. Some predictability in meeting form usually makes meetings more effective in exchanging information and solving problems.

- *How will the discussion occur?* Anyone who has participated in total quality management is usually familiar with teams writing and using *ground rules* to govern their discussions. The group makes up a list of the rules it wants to use to ensure that discussions are productive, creative, and worthwhile.

  Typical ground rules that project teams devise include the following:

  One person talks at a time.

  Everyone listens while one person speaks.

  Treat disagreements with respect.

  In discussion, the group goes in sequence.

  No new idea is a bad one.

  Once a group has devised its ground rules, it specifies how to implement them. For example, some groups that struggle with members' punctuality impose a rule that late arrivers must buy donuts for the next meeting. Others demand that late arrivers contribute to a lottery fund or charity at the rate of one dollar for every minute they are late.

- *Is the meeting leadership effective?* The leader's ability to run meetings is crucial to the overall effectiveness of the meeting. Most of the people who lead project meetings, however, are not nearly as skilled at this task as they are at the more technical sides of their work.

  The first step to enhancing leadership effectiveness in project meetings is to have the project team spell out the tasks of the leader: What specifically must the leader do to make the meeting most effective? If the group does its job well, it will specify tasks not only for presenting information clearly but also for organizing discussion to achieve full participation. Typical meeting leadership tasks include the following:

Make sure everyone who can contribute to an issue is given a full opportunity to speak out.

Don't allow one or two people to dominate a meeting.

Call on the quiet participants; they often have much to contribute.

Post and follow an agenda.

Circulate the agenda before the meeting so people can think about the key issues.

Put as much effort into organizing a useful discussion as you do into presenting information.

Such a listing of leadership tasks helps the leader to clarify what he or she has to do in order to be effective. The list makes it easier for the leader to receive and act on feedback on his or her performance because people comment not on the leader's personality but on specific behaviors that can be learned and changed.

Once leadership tasks are identified, it also becomes possible to assign some of them to other project team members, or even to share leadership on a rotating basis. It is usually efficient to assign leading meetings to one person, but not always effective.

Many times, the person who is in charge of the project will want to lead meetings in order to be in control of the meeting's content and process. Suggesting that members of the team share or rotate leadership may feel like a potential loss of control, an inefficiency, perhaps even an invitation for chaos.

Sharing or rotating leadership, however, makes participants more active not only in the discussion but in the structuring of the meeting. Rotating leadership makes it more difficult for people to take a passive stance towards discussion and creates a sense of ownership for the meeting's success.

Daily Project Meetings

Just as the daily production meeting has increasingly become a standard practice in manufacturing, daily project meetings are becoming standard on many projects. The project team uses the daily meeting more for quick updates than for the more in-depth kind of problem solving that occurs in weekly meetings. People usually hold the daily meeting in the trailer, on site, or after an initial walk through of the site. In order to keep the meeting brief and to the point, participants may stand.

Some people resist the idea of a daily meeting because it strikes them as just "one more meeting," a waste of time. Yet a scheduled meeting, especially if it is brief, can resolve a great many concerns in a short period of time. The fact that the meeting is scheduled also enhances its effectiveness. Whenever there is no scheduled meeting, people can spend hours simply trying to find each other on a given day. The brief morning check-in enables people to inform each other of their whereabouts so that there are no obstacles to making decisions or solving problems.

The most successful daily meetings we have seen use a flip chart on an easel to record, in a very simple way, the issues of the day (see Table 12.1). The chart is divided into two columns, *issues* and *actions*. Each person lists whatever issues he or she thinks the groups should review. When an item is discussed, it results in some kind of action step. The page is left in the trailer until the next day, when completed items are crossed off and incomplete items are carried forward.

By using a flip chart, participants in the meeting eliminate the need to work with time-consuming minutes, yet they produce a record of the discussion that strengthens follow-through activities on the items that get discussed. In addition, the flip chart is visible and available to any other project team member who cares to look at it. People who themselves do not attend the meeting can still find out what occurred by checking in with the flip chart.

Decision-Maker Meetings

Problems such as the increasing project RFI backlog, the dispute between the carpenters and the contractor, the "surprise" conditions on the site that everyone is going to have to adapt to, the increase in materials cost, and the bankruptcy of a key subcontractor may arise at any time. How can the project team get answers to particularly difficult problems without scheduling an additional partnering workshop or calling a lawyer? Where do people working on the project turn when they need an answer and when they do not have the authority to provide the answer themselves?

Table 12.1 Daily Check-In Meeting
(Sample Items from Meeting Flip Chart)

Issue	Action: who, what, when, where, etc.
HVAC workers are complaining that electrical workers are getting in their way.	HVAC supervisor and electrical contractor talk today at 9:00 A.M., meet with combined group at noon, and write joint memo to both groups.
Plumbing expects a delay in work scheduled for this week due to materials shortage and supply problems from manufacturer.	General contractor reschedules plumbing and meets with all other subcontractors today at 10:00 A.M. to discuss how they will adjust to fill in gaps left by plumbing delay.
Carpenter has an idea to improve fittings on interior work.	General contractor calls the architect to discuss idea, involve carpenter if needed.
Carpenters say they have heard that the sheetrock subcontractor is experiencing financial troubles and may go bankrupt.	.General contractor calls the sheetrock subcontractor, and also calls three other sheetrock people as insurance.
Electrical workers are coming in ahead of schedules, under budget.	Owner writes note of thanks by the end of the week and orders in pizza for rest of week.
Town inspector is coming tomorrow.	Prepare areas being inspected.

One of the most effective mechanisms for managing project communications overall is a monthly project decision-makers meeting. Usually attendance at this meeting is limited to just a handful of people, and often this handful does not include the people who work on the project on an everyday basis. Thus, for example, the group might include the owner of the contracting and key subcontracting firms (but not the site superintendent), the architect's principal in charge of the project (but not the project architect), and the client (but not the client's project liaison).

The meeting is scheduled well ahead of time and may occur at a site that is neutral to all the members of the group. Many people prefer a restaurant setting and dinner meeting for this session to set a tone of cooperation and building relationships. There may be a recurring agenda, but the main point of this meeting is to provide a proactive forum to answer and resolve the inevitable questions and ambiguities that emerge on a project and that need responses from this group.

Observes an architectural firm owner,

> Having this kind of meeting scheduled works for us in two ways. First, it establishes a foundation so we can work out issues ourselves. By the time we have our third or fourth meeting, there is usually something of substance we have to talk about and resolve. At that point, we're all grateful that it's not our first meeting.
>
> Beyond just us, the meeting also creates a sense of security on the project. Because my people on the project know there's a place where their questions will get answered, it encourages them to settle more issues on their own.
>
> I guess the other important thing is that our meeting together sets a tone for the project. It's a little harder for our people to be fighting with each other when they know we're going to be sitting down for dinner every month.

Informal Meetings. The more effective the weekly and daily meetings are, the less need there will be for informal meetings. Still, new issues will always arise, and people will need to get together quickly, usually to solve problems, share information, and make decisions.

While it is not possible to schedule informal meetings, it is possible to address the two most common problems people encounter in trying to make these meetings effective: access (peoples' ability to find each other when they need to) and context (how the meeting fits within the larger context of the project).

The chief problem with access is a Murphy's Law kind of situation in which people never seem to be available when someone needs to find them. This is a problem in any kind of business, but a particularly difficult problem in design and construction because the workplace is usually large enough to get lost in. In addition, key people on any project may be working on several different sites at the same time.

We have seen people most successfully address the problem of access with the two major strategies for managing communications that provide a consistent theme throughout this book: scheduling regular meetings ahead of time and providing backup means of access for times when the conventional methods fail or fall short.

Scheduling Regular Meetings among Key Pairs and Trios of Project Team Members. As with weekly and daily project meetings, key pairs or trios of project team members will often realize that it is a useful thing for them to check in with each other on a regular basis. This is often a useful topic to include on a part-nering workshop agenda, to ask the project team members to think about key pairs and trios of people that might benefit from getting together on a regular basis. They try to anticipate the fre-quency with which they need to do this and schedule a pre-dictable meeting, often at the same time each week.

For example, a general contractor and a key subcontractor might schedule a weekly lunch with each other at times in a pro-ject when the subcontractor's work is especially difficult. Or the architect and the general contractor might schedule a weekly evening beer as a way to provide access to each other before they encounter one another in the more tense atmosphere of a weekly project meeting.

Scheduled meetings do a great deal to solve the problem of access. People already know they are going to meet, and they have made time available. Many times when people panic because they need to discuss an issue with another project team member, the panic is due to a fear that it will not be possible to find the other person, not to the difficulty of the problem that must be solved.

Anticipating a regular, scheduled meeting, people may accu-mulate a list of discussion items over the course of the week. Thus the meeting becomes an efficient means to resolve several issues at the same time and to connect solutions on one issue to solutions on the others. If there are no current problems to discuss, the meet-ing can always be cut back or even eliminated for that week, but at least the forum is there if it is needed.

Improving Meeting Effectiveness

Many people have had bad experiences with meetings that are not productive. Thus they may be reluctant to support the use of meet-ings of any kind for project management and communications.

At the same time, design and construction project team members need some way of exchanging information, solving problems, and making decisions. Often a meeting of some kind can efficiently and effectively provide these functions if it is well run. In partnering workshops, we put project team members to work to devise several different kinds of productive meeting plans and procedures.

We emphasize this aspect of partnering because when it comes to meetings of any kind, we have seen several principles apply as follows (see Fig. 12.2):

1. *The best meeting procedures start with a clear purpose.* Focus on the information exchanges that need to occur, and the optimum meeting procedures will follow.

In the more than 250 team-building and partnering assignments we have worked on, we have found that most teams don't use meetings effectively. They either meet too often or not enough. When they do meet, they don't get much done. They try to use meetings to accomplish tasks better tackled by other tools (e-mail, fax, etc.) and they fail to use meetings to accomplish what they could.

Because meetings are tangible, people often begin a discussion of *communications procedures* with a discussion of meetings specifics: times, dates, places, etc. This is understandable but problematic

---

1. The best meeting procedures start with a clear purpose. Focus on the information exchanges that need to happen and the optimum meeting design will follow.

2. Scheduled meetings work better than "as-needed" ones.

3. Short, frequent meetings get more done than long, infrequent ones.

4. Flip charts provide an effective tool for record keeping.

5. Brief, regular feedback helps keep meetings productive.

6. Meetings are two-way events.

7. For productive discussions, break the group into subgroups.

Figure 12.2. Improving meeting effectiveness: Principles to improve project meetings.

because it will often quickly result in disagreement; people argue over methods because they have failed to reach an agreement about the meeting's purpose. Arguing over meeting specifics can pose a much smaller problem for the project, however, than agreeing on specifics and settling for meetings that are efficient but not effective.

The place to begin a discussion of any meeting procedures is not trying to agree on a time and place to meet but rather to examine the key information exchanges that need to take place and to define the project tasks that are best accomplished (or perhaps that can be worked on only) by a group (see Table 12.2).

Working on these two topics may take between 30 minutes and an hour of discussion overall in a partnering workshop. Defining information exchanges and communications tasks begins by asking each individual at the workshop to identify what kinds of information he or she needs from every other project team member, along with an approximate frequency for how often the information needs to be updated. Participants also list who they need to interact with for other purposes aside from exchanging information: making decisions, solving problems, etc.

The whole discussion is logged on a flip chart. The log provides a visual record of information exchange content and frequency. From this log it is possible to ask the group to begin to map out meeting schedules and topic agendas that match the group's needs.

Usually when design and construction project teams go through this brief bit of work they discover that the original plans they had for meetings would not have accomplished the informational tasks that need to be addressed.

2. *Scheduled meetings work better than "as-needed" ones.* Some people resist scheduled meetings of any kind simply because they "don't like meetings." Yet when a meeting is scheduled on a regular basis (e.g., a weekly project meeting or a brief daily "check-in" meeting of key team members), the scheduling in itself takes care of the major problem of finding a time that works well for everyone. In addition, scheduled meetings make it possible for people to accumulate their own lists of items to discuss so the meeting can efficiently resolve several issues.

Scheduling meetings also makes them more useful as a planning tool. When meetings are held on an as-needed basis, then

Table 12.2 Information Exchanges and Tasks: Sample Matrix Items

| Person receiving information | Information needs | | | Tasks to be accomplished |
	Provider	Content	Frequency	
General contractor	Architect	Ongoing clarification of design intent and work-ing drawings	Weekly	Answer questions so that construction can proceed
General contractor	All subcontractors	Update on current schedules and specific problems	Daily	Coordinate schedules and solve problems
General contractor	Electrical worker	Update on current work using new tools and technologies	Twice a week	Stay on top of new methods and tech-niques
Architect	General contractor	Review progress of con-struction	Weekly	Answer and resolve questions and ambiguities
Electrical worker	General contractor	Get guidance for sched-uling and solving problems	Daily	Resolve issues
Electrical worker	All other subcontractors	Find out about others' work as it may affect me	Weekly in general plus daily updates	Anticipate emerging issues and problems
Owner	Project team	Get cost data in a timely way	Weekly	Find and limit overruns and delays
Building manager	Project team	Make sure specific deci-sions keep me in mind	Weekly	Keep the building occupants in the picture

they usually occur in a state of crisis or emergency, with immediate conflicts or problems to solve. Scheduled meetings contribute to a more proactive, organized project environment.

3. *Frequent, short meetings work better than infrequent, long ones.* Monthly project meetings may provide interesting stories and bring people together but it is usually hard for them to be the place where much work gets done.

Because things happen quickly on a design and construction project, and since delays cost time and money, meetings that are scheduled frequently tend to get more done than meetings that are scheduled infrequently. Also, when meetings are scheduled frequently, they don't need to take so long because they provide a timely point of entry for resolving emerging problems.

4. *Flip charts provide a useful tool for note keeping.* Using flip charts to record discussions provides an efficient and visible means for organizing discussions while they are occurring. Writing the key points of a discussion helps organize and focus the discussion while it is happening and provides clear documentation of the discussion for later reference. In fact, flip charts often eliminate the need for the time-consuming recording and circulation of meeting minutes. Flip charts also provide a brief visual record of the meeting for people who did not attend.

5. *Brief regular feedback helps keep meetings productive.* Meetings of any kind can lose focus, energy, and value. However, if participants are given a chance to discuss their perceptions of which aspects of the meeting are working well and which need improvement, they usually provide the information needed to bring the meeting back on course. This feedback can be collected in writing in brief, informal comments or in a brief discussion.

It is usually not necessary to collect and work with feedback at the end of every meeting. Often, checking in for feedback at every fourth or fifth meeting provides sufficient information to keep the meetings on track.

6. *Meetings are two-way events.* One reason people dislike meetings is that they have often attended many one-way meetings, i.e., meetings where someone made an announcement with little attention paid to a discussion afterwards. This kind of one-way meeting accomplishes just half of the meeting's work, presenting the information.

The other half of the meeting, discussing the information, requires careful planning in order to be effective. Quiet people who have a great deal to contribute may be left out because they are uncomfortable participating in the discussion. People who like to speak out may dominate.

7. *For productive discussions, break the group into subgroups.* Many of the kinds of problem solving project teams attempt to do can be done most effectively by taking the group and breaking it into subgroups. Conducting a discussion in this manner has several advantages:

- *It equalizes participation.* It is easy to not speak out in a group of eight, ten, or twelve, but very difficult to not speak out in a group of three.
- *It provides for depth of discussion.* While larger groups can contribute more ideas in the idea generation or brainstorming parts of problem solving, they can become cumbersome when trying to take the ideas and develop them. Developing an idea means working more closely with it, refining it, and moving it along to more advanced kinds of refinement. Such movement is difficult under any conditions, but much easier with a group of three than with a group of seven or eight.
- *It increases focus and control.* Participants at project meetings tend to have very definite ideas and opinions, so it is easy for the meetings to lose focus and control. Moving people into subgroups makes it more difficult to disrupt or derail the meeting. There is too much movement and focus for potential disrupters to take aim.
- *It involves several group members in leading the meeting or rotates leadership.* In order to stay on track, meetings need leadership, but they don't necessarily need one leader.

Experienced leaders know they are going to have difficulties in a meeting when people who attend tell them, "I'm here for your meeting." The leader will work against this imbalance of ownership and responsibility for the whole meeting and after the meeting when it comes to implementing decisions.

The more the tasks of leadership are broken down and assigned to several team members, the more ownership the other team members have for the meeting's success. Other group members can lead a segment of the meeting or organize a small discussion. If the leader's goal is to spread a sense of ownership for the meeting's results, it can be more effective to have team members lead a discussion rather than simply present more information. Leading a discussion puts them more in the driver's seat of responsibility for involving others and making sure that the discussion is balanced and productive.

First Person: Participant Comments about Better Meetings

We spend so much time designing buildings. Why not design meetings too?

Project Architect

When we were working on all the meetings procedures, I thought, "This is a waste of time." Once we finished, though, I could see that by spelling things out, we're going to make all of our meetings a lot more worthwhile.

Electrical Contractor

I like to think of myself as a pretty good communicator, but working on all these procedures reminded me of how much I often leave out. It was kind of difficult to work on, but I can see that the success of the project depends as much on my ability to communicate the design as on my ability to conceptualize the design.

Project Architect

I've been involved in design work before when people tried to be nice to me, but they never really paid attention. Now that we have clear procedures for meetings, I can see that it will be a lot easier for me to represent my concerns and be heard by the project team.

Building Manager

13

Electronic Partnering?

Using Groupware Computer Software to Enhance Design and Construction Partnering*

Summary

Groupware is computer software that can solve many of the communications problems in design and construction work. Groupware can enhance partnering substantially by providing the means for people on a design and construction project to:

- Communicate with each other quickly and easily
- Log and document all communications
- Open communications between pairs or small groups of people to all members of the project team

At first look, many people think groupware could replace partnering as it produces, "automatically," many of the same results as partnering efforts. On closer review, however, it becomes clear that it really works in the opposite way. Partnering is necessary to ensure that groupware is used fully and properly.

*This chapter is based on a case study the author prepared for research conducted by the Space and Organization Research Group in the College of Architecture at the Massachusetts Institute of Technology; the research was funded by the Takenaka Corporation of Japan.

What Groupware Is

Groupware is computer software that enables people working together to communicate with each other to share information, examine complex databases, and work in teams simply by reviewing and sharing information on their computers. All kinds of businesses are discovering uses for groupware, finding that it has widespread applications and substantial returns and that it enhances existing processes and creates possibilities for new ways of interacting.

Lotus Notes is one of a half-dozen kinds of groupware offered by different manufacturers. This chapter focuses on Lotus Notes because it is currently by far the most widely used kind of groupware. A *USA Today* article in May 24, 1994 claimed over 750,000 Lotus Notes users in over 3,200 companies. By December, according to an article in *Fortune* (December 12, 1994), the numbers had risen to over 1 million users in over 4,000 companies. The *Fortune* article also points out that Lotus Notes sales for 1994 of 600,000 copies are doubling annually.

Lotus Notes enables everyone on a team to share information easily, in ways that differ significantly from electronic mail and faxing. People on a Lotus Notes network use Notes to send messages to each other. If they wish, can leave those messages open and visible to any or all of the other members of the team.

The aspects of groupware which make it appealing and impactful for design and construction partnering are as follows:

- *User friendly.* With its point-and-click use, groupware attracts many who shy away from more complicated computer applications.

- *Inexpensive.* At about $400 to $500 per seat, groupware costs about the same as many popular spreadsheet and word processing programs.

- *Technologically simple.* System requirements are minimal. Users need a current computer or laptop, a basic local area network or modem, and a server or large capacity hard drive to store information.

- *Flexible.* Users can easily tailor and customize groupware to develop their own menu of options and applications.

- *Entertaining.* By providing access to information previously unavailable to users, groupware piques curiosity and stimulates creativity.

Groupware does for communications what word processing software does for writing: it makes it possible to get much more done, more effectively, with no more effort. Groupware applies the power of a computer to the tasks of exchanging information among the members of a project team.

These attributes may seem straightforward and simple but they can easily add up to impressive results. For example:

- Coopers and Lybrand report using Lotus Notes to network employees in five cities and three departments to produce and deliver a proposal overnight, winning a multimillion dollar consulting contract.

- London-based Triangle software discovered that Lotus Notes kept employees well enough connected to the company as to render unnecessary a $150,000 per month office rental expense. With the groupware, more people could work at and communicate effectively from their homes.

- Ann Palermo, a technology consultant at International Data Group in Framingham, Mass., comanaged a study of Lotus Notes impact on productivity in a collection of businesses. She charted a 179 percent return on investment in Lotus Notes in the third year after installation. For service businesses, she recorded a 351 percent return.

- Police departments are scanning in criminal photographs and fingerprints to exchange data easily across state lines and with communities.

Groupware Hardware Requirements

Lotus Notes can be purchased in single or multiple copies at a retail software store such as Egghead or Softpro. More typically, however, companies purchase groupware through a value added reseller (VAR). VARs sell groupware as a product in itself, but their chief business is to provide integrated services including:

- Installing groupware on existing computer and network hardware

- Designing forms and databases for the specific uses of the group
- Training users in the basics of using the groupware
- Debugging the system overall to make it work

Lotus Notes needs a large hard drive (called a server) to house the data shared on the network. The server can function with as small as 500 MB, but 1 kB is more typical. (These days the street price for a drive of this size can be as low as $1000. Drawing from the data on the server, Lotus recommends that each computer connected to the system use at least a 486 processor. Lotus Notes will run on a 386 processor, but it will be slower.

Computers can be connected to the server with network hardware such as local area networks (LANs) or wide area networks (WANs). It is also entirely possible to connect to the server without any wired network, just by using a modem. This ability of groupware makes it very appealing for applications where the members of a group are in different locations.

What Groupware Can Do

Groupware enables project team members to send electronically the same kinds of communications to each other as they usually do on any project. The differences are that with groupware:

- The communication is instantaneous
- The communication is logged and documented on the server and on a list visible to all project team members
- The communication is kept open for other project team members to see if project team members wish

It is possible to customize groupware for specific applications. The examples cited in this chapter were developed in a pilot application of Lotus Notes for design and construction in the building of Lotus' corporate headquarters in Cambridge, Mass. (This case is referred to as "Rogers Street," the location of the building, in the rest of this chapter.) Although the applications are typical of those useful in many design and construction projects, it is important to keep in mind both that it is not necessary to use these applications and that it is possible to develop other applications fairly easily.

To work with groupware, a project team member uses three aspects of it:

- *Database.* The database used most often on a project is usually the directory, the list of names, and addresses, phone numbers, and fax numbers of project team members.
- *Forms.* These are templates of the types of communications used most often in project communications (see Fig. 13.1). The Rogers Street building pilot used 11 different forms:

  1. *Action items.* Priority actions listed by person. These enable people to review what others are working on.
  2. *Applications for payment.* These messages did not list the amount of a payment, but simply recorded that the work was completed. This helped people to know when various phases of the job were complete.
  3. *Change orders.* These are official decisions to alter or modify the original design. They are likely to impact many people working on the project. The log of change orders is an important record, as it tracks alterations that may impact design and costs.
  4. *Fax cover sheet.* This provides the sender of the fax with a template to send a fax to anyone in the directory electronically, as well as to people not in the directory through their fax machines.
  5. *Daily reports.* Several of the members of the project team logged brief daily reports of their activities.
  6. *Memos.* This is an informal form, simply listing the date, "From," "To," and "Subject." Perhaps because of its simplicity, people on the project team in the pilot case used this form more than any other for most of their everyday communications.
  7. *Meeting minutes.* These are kept on groupware so that people can easily refer back to decisions and agreements made in meetings.
  8. *Meeting agendas.* These are put on groupware before a meeting so that people can add items on their own and so that those attending a meeting can know, well before the meeting, what is to be done.

1. Action Items

2. Applications for Payment

3. Change Orders

4. Fax Cover Sheet

5. Daily Reports

6. Memos

7. Meetings Minutes

8. Meeting Agendas

9. Requests for Information

10. Requests for Proposals

11. What's New

(This is what project team members see when they double-click on the "Forms" heading of a Lotus Notes application designed for design and construction project applications. In this sample, the team member has selected the "Memos" option prior to preparing a memo to send to another project team member.)

Figure 13.1.   Types of communications for which templates are stored in the computer.

9. *Requests for information.* This is the official American Institute of Architects' form for requests for information. The purpose of the form is in part to provide the needed information, but also to track and document the exchange of the request.

10. *Requests for proposals.* This is the official form that the project uses to solicit bids on special tasks. These may include work beyond the original scope imagined or highly specialized tasks that are outside the experience and expertise of any of the members of the project team.

11. *What's new.* This is designed as a more informal, random bulletin board for project team members to communicate on current concerns.

These eleven forms are displayed as a simple list under the word "Forms," which appears as a heading on the top of project team members' computer screen.

Using Groupware: Sending Communications

Project communication of any kind begins with the project directory, a simple database including the names, firm names, addresses, and phone and fax numbers of all the people on the project team. One uses Lotus Notes by first opening the groupware program on the computer. The first screen the program shows is a listing of the forms or types of correspondence (see Fig. 13.1). At this point it is possible either to send a communication or to review the running log of communications of a particular type.

To send a communication it is necessary to compose a document. This involves clicking in the appropriate places so that the computer provides a template form for the type of correspondence chosen. For example, to send a fax, one clicks on "Fax" and the computer displays a fax transmittal form. In the sample chart of forms, the project team member has clicked on "Memos."

Once a particular form is opened up with a double click, it displays a template prompting the most common headings. Each heading is followed by a blank space so that the project team member can type in the appropriate information. At appropriate headings, the computer supplies a list of the whole project team so that the person composing the form can simply select the person without having to key in the person's address, phone, etc. Once the person has been selected, the computer can supply all the necessary information about sending the form by fax, by e-mail, or just over the network. For a sample memo form, see Fig. 13.2.

The ability to send memos, faxes, and correspondence directly from the computer screen is itself very helpful. The sender doesn't have to go through the time-consuming steps of printing the file from the computer and then standing by a fax machine. The receiver doesn't have to deal with curled paper. In fact, paper does not have to enter into the transaction at all.

It is also possible to send any correspondence or fax to as many people on the directory as desired simply by highlighting those peoples' names at the top of the list. This feature makes it very easy for people who are communicating directly with each other to provide access to the information they are exchanging to others who might benefit from knowing about it.

This feature actually leads the sender to consider who should receive the communication, as it lists on the fax or correspondence form the names of everyone involved in the project. The sender must go through the list in order to find the names of the people he or she wants to receive the correspondence. Going through the list entails thinking about everyone who is listed.

Using Groupware: Reviewing Communications

To review others' documents one clicks on the type of document to review and the computer displays a running log of who sent what to whom. One can review who sent whom faxes, change orders, memos, requests for information, requests for proposals, etc. If one sees anything that looks relevant (or even just interesting) one can click on the particular document and the computer will show the content of what was sent.

For example, a contractor might be interested in correspondence between the architect and the engineers because such cor-

Memos

(Sample of "Memo" form template used to compose informal memos)

Date:

From:

To: (When the person using the groupware clicks on this line, the computer displays a list of project team members so the person using the groupware can simply click on the appropriate names.)

Subject:

When reply is needed:

Message:

(The groupware supplies the headings for a memo. The person composing the memo keys in the appropriate information. The groupware sends the memo to whomever the sender lists at the "To:" heading.

Figure 13.2.   Memo template.

respondence might be about changes in the building that will affect him. An the architect might be curious about correspondence between the contractor and a particular subcontractor on an aspect of the design that might be difficult to work on. The developer might be interested in any correspondence between the architect and the owner. The owner might be interested in all the correspondence in the log.

The log in Fig. 13.3 is a simulation of a typical listing of memo-type communications that occur on a project.

How Project Team Members Use Groupware

In the Rogers Street case, as well as in several other pilots studied more briefly, people on the project team use groupware in four major ways:

1. *Groupware becomes part of everyone's everyday routine.* The contractor, owner, architect, developer, and interior designer all have a daily routine that begins with a visit to their computer. They log on and review the list of who sent what to whom on the previous day. On a typical day, they might see between 10 and 15 transmittals on the list. If any of the transmittal headlines interests them for any reason, they double-click on the headline and then look into the complete correspondence.

2. *People use the memo form.* Despite all the possible forms for communications that reside in the computer, most people use the simple memo form most often. The reason isn't that the other forms are terribly complex, it's more that the simple memo gets the job done.

3. *People use the fax capability extensively.* All of the project team members say they like the "instant" fax capability groupware offers. They say that eliminating the need to use the fax machine makes it much easier to get communications out to the other members of the project team.

4. *Project communications are noticeably relatively limited in the areas groupware is not used.* The groupware application did not include the full team of consultants on the pilot project. At the outset, that did not seem to cause a large problem, but the gap in their

Date	From	To	Subject
5-17	On-site Const Mgr	Project Architect	Level of detail in drawings for 3d floor electrical work
5-17	Interiors Arch	Project Architect	Special materials for special 2d floor conference room
5-17	Construction Mgr	On-site Const Mgr	Delay in shipment of window frames and materials
5-17	Developer	Project Architect	Prepare for meeting with City Inspectors
5-17	Owner	All	Holiday schedule upcoming, need to coordinate
5-17	Owner	Electrical Engineer	Details of needs for electrical in computer facilities
5-17	Project Architect	On-Site Const Mgr	Level of detail in drawings for 3d floor electrical work
5-17	Owner	Interiors Arch	Clarify special materials for special 2d floor conference room
5-17	Plumbing Engineer	On-Site Const Mgr	Request to schedule 3d floor work before carpenters start
5-17	Electrical Engineer	On-site Const Mgr	Review approvals needed for 3d floor electrical work
5-18	Telecom Engr	Proj Arch	New technology available for use in project
5-18	Owner	All	Reminder for upcoming project meeting
5-18	Project Architect	Const Mgr	Request more details about delay in window material

(The 5-17 entry from Owner to Electrical Engineer is highlighted, ready to open up and review.)

Figure 13.3. Sample project memo log. This is a replica of the log of project memos as it would appear on a computer using groupware.

understanding and knowledge of the project became noticeable six months into the project.

People in the pilot program did not use groupware to communicate about contract amounts; that area was thought to be confidential at the outset. This has also created a noticeable gap in information people have about the project, as most people are very informed on other matters.

Groupware Impact on Project Communications: The (Not So) Basics

Groupware's major impact is in improving basic everyday project communications. Specifically, groupware:

- Provides project team members with more information they need more quickly and more completely than any manual system
- Reduces miscommunications, "lost" documents, and conflict
- Improves the quality and quantity of information available to all members of the project team
- Reduces the amount of time and effort project team members must devote to organizing, locating, and sending communications
- Improves documentation, resulting in more efficient communications

A project team member explains, "People don't make excuses for not getting information. They know they have it. Also, if they really do lose it, they can very easily go back into the computer file and make themselves another copy."

If the groupware did nothing else, these "basic" accomplishments alone would be significant because of the influence communications has in design and construction. If everyday communications are improved, then the costs and conflicts stemming from communications problems will be reduced. If communications are easier to transmit, receive, and manage, then the learning and creativity that must take place in the problem solving that is essential in design and construction should occur with greater ease and effectiveness.

Groupware Impact: Beyond the Basics for Individuals

In addition to improving basic communications on the project, groupware also has the potential to enrich and expand overall design communications from the perspective of both individuals and organizations.

For individuals, groupware can:

1. *Help them to be more* active *in design communications.* More people communicate more often and more easily because groupware makes it easy.

2. *Encourage them to be more* universal *in their stance.* Because groupware provides project team members with information outside their usual frame of reference, it makes it more difficult for them to maintain an individual scope of participation, focused only on their own position.

To better understand the first and second points here it is useful to use Table 13.1, which outlines the design communications process along the axes of active–passive and universal–individual. Problems in communications arise when people on the project team are passive, or when they are active but the focus of their activity is from an individual stance or focus. When people are both active and universal in their scope of interest, the quality of the design communications and problem-solving process improve. Using groupware provides project team members with information that moves them toward block 1, more active and more universal in participation.

3. *Groupware increases project team members' overall scope of influence and participation in the design process.* One simple but recurring problem in this area is that architects complete their phase of design work months before it is built. Then, when contractors are working on the job, the designer who conceptualized what they are constructing is out of the information loop of everyday project activities. Contractors improvise solutions to questions they have

Table 13.1 Participation in the Design Process

Level of activity	Focus of participation	
	Universal	Individual
Active	Active + universal Block 1	Active + individual Block 2
Passive	Passive + universal Block 3	Passive + individual Block 4

about the drawings, and the result is a finished product that does not resemble what the designer had in mind.

Groupware helps remedy this problem by providing information to people when they are outside the traditional information loops. In the pilot case, several people whose work had been formally completed for many months kept in touch with the project by reviewing project communications logs. When they saw information relevant to them, they were able to make sure their input was heard.

Groupware Impact: Beyond the Basics for Organizations

Beyond its impact for individuals, groupware also influences the firms involved in the project.

1. *Groupware brings more information on the project to the whole organization.* This enables people usually outside project information loops to improve their own work and contribute to the project. For example, draftspeople not directly connected to a project can see when the project runs into delays and adjust their work priorities accordingly.

2. *Groupware brings internal communications system gaps and problems to everyone's attention.* People on a project team make commitments to each other to process information, answer questions, solve problems, and clear up ambiguities. Often the person who responds to a question must rely on another person in the firm to supply the response, but the other person may fail to follow through effectively.

It is easy for the person in the firm to blame people outside the firm for these miscommunications, but it is more difficult to do so when the communications outside the firm are chronicled on a computer log. Working with groupware pinpoints a firm's internal communications gaps.

3. *Some firms develop their own groupware applications for other uses.* In addition to project work, groupware has numerous other potential applications in a firm: managing projects overseas, tracking expenses, and keeping new technologies bulletin boards.

Getting groupware in-house for a project means having it available for people to develop these applications if they wish.

Groupware's Potential for Partnering

Groupware's potential impacts for communications, for individuals, and for organizations parallel the impacts of partnering efforts. Without groupware, partnering programs usually attempt to improve communications. Partnering can result in impacts similar to those of groupware's for individuals' participation in the design process and for improving firms' internal communications processes. Thus groupware can reinforce what partnering often attempts to achieve.

Groupware also has the potential to provide an infrastructure for many partnering tasks, a useful means to accomplish much of the work people do when they partner. More specifically, groupware can impact partnering by:

- Providing a specific tool to develop communications procedures
- Enhancing communications effectiveness, thus causing people to set partnering goals higher
- Providing documentation that can assist in resolving problems
- Enhancing the ability of individuals and firms participating in partnering to process information

Achieving these potentials may not be as easy as groupware can make it look, however. Implementing groupware so that it enhances partnering means understanding both the difficulties involved in implementing new technologies in general and the limitations of the groupware.

External Limits to Groupware's Effectiveness

Although groupware has the potential to achieve the improvements described above, both external and internal factors can limit its impact and effectiveness. The external factors are outside groupware in the nature of technological change and in the design and construction industry. The internal factors are inherent in the groupware itself.

1. *Resistance to change.* Implementing groupware effectively in design and construction means implementing an innovative, new technology. On the surface, this may not appear to be a problem. The sample memo logs and fax forms listed earlier in this chapter do not appear to depart radically from the way people in design and construction already do business. However, using groupware involves change, and the track record for successfully implementing technological change of any kind is not very promising.

There is an extensive research literature on technological change, which is largely a literature of failure and frustration pivoting on one key word: resistance. Historically, even the most apparently useful and beneficial new technologies were resisted and delayed. It took years for the process of pasteurization to win approval, decades for the Navy to use steam-powered boats. More currently, any organization that attempts to computerize or change operating systems runs headlong into peoples' deep-seated impulses to resist change.

Groupware provides a simple tool but its use in project communications is not simple at all. The full adaptation of groupware in design and construction would also require several major changes in the ways people involved in the process communicate with one another and think about their own work.

The literature on organizational change implies that however strong its potential, the full implementation of groupware will not be a smooth and easy process. Even though it is easy to use, inexpensive, and seemingly in the interest of everyone to apply it fully, groupware is likely to follow the pattern of other apparently useful technologies and fall short of expectations.

2. *Limited computer applications in design and construction.* While the design and construction industry uses computers extensively for computer-assisted design (CADD) applications and for some cost and word processing work, other kinds of businesses (banking, manufacturing, etc.) use computers more extensively. Most middle managers in insurance companies use their computers for everyday correspondence, calculations, and communications. Many mid-level design and construction professionals, however, still don't have computers on their desks. Though their firms may

use complex computer-assisted software, most of the people who use the equipment are either lower-level draftspersons or technologically minded "CAD jockeys." This lag in computer applications is reflected in groupware, which is used much more in other businesses than it is in design and construction.

The lag in computer applications overall makes it just a bit more difficult for people in design and construction to use groupware. There are still people who are unfamiliar with using Windows, working with a mouse, and using a keyboard. These may seem like minor problems, but they can greatly curtail the use of groupware.

3. *Fragmentation of the industry.* In other businesses, groupware usually connects people who have an interest and/or a track record in working together. In design and construction, groupware attempts to link people who are not used to being linked, whose professions, trades, and companies may be better organized to reward competition than collaboration.

4. *Interdependence.* Although the design and construction business is highly fragmented, the various firms and organizations on any specific project are highly interdependent. This need to interact, coupled with the fragmentation of the industry, places a high amount of pressure and tension on everyday project communications and would similarly impact groupware.

Inherent Groupware Limitations and Potential Problems

When some people first encounter groupware, their first reaction is that it has the potential to replace partnering. On first review, groupware seems to do "automatically" what people struggle to do in partnering workshops. From the cases we have studied, however, the process works the opposite way.

> Aside from groupware not being able to replace partnering, partnering is necessary to implement groupware effectively and fully.

In order to understand why this is true, it is worth considering what groupware can and cannot influence. This involves exploring not the external factors that limit groupware, but the limits inherent in the groupware itself.

Groupware provides a tool to improve communications, but it does not necessarily impact what partnering focuses on: how project team members use the tool. A contractor involved in the pilot case summed up the situation, "If you're ornery before you get the groupware, you'll still be ornery after you have it." Some more specific distinctions between what groupware does and does not do are listed in Table 13.2.

The right side of the table, "What groupware does not necessarily influence," involves three different issues: full use, "fit," and the intentions with which people use groupware.

1. *Full use.* One problem with groupware is that it is not possible to tell if people are using it fully. The sample project memo log in Fig. 13.3 makes it look like everyone on the project is using groupware for all communications. However, the log only documents what people put into it. It has no way of showing when they use the phone because they want to keep an awkward communication off the network, or when they refrain from communicating at all because they don't like to type on a keyboard.

"Full use" of groupware is subtle and intangible, and is difficult to document and assess. At the same time, it is crucial for groupware success. If all the people on the project team fail to use groupware fully, it will fail to attain the lofty potentials outlined above. Moreover, with less than full use, groupware could actually create quite misleading impressions that everything that needs to be communicated on a project has been resolved.

2. *Fit.* A second set of the limits of groupware has to do with the issue of "fit," i.e., whether it fits with the way individuals and firms do business (see items 2 and 6 in Table 13.2). If the fit is close, there is no problem, but if it is not, people might not use groupware extensively enough so that it can work as a projectwide tool.

Individual fit involves integrating groupware into the way people do their jobs. For people used to working with computers in any way, adding groupware is usually not at all difficult. Many people in design and construction, however, do not use computers much for their work. Some contractors still prefer manual methods for the math work they need to do. Some architects and engineers use computers for CADD applications, but are uncom-

Table 13.2   Groupware Limitations

What groupware can do	What groupware does not necessarily influence
1. Improve 1-on-1 communications among team members	1. Whether people use it
2. Make 1-on-1 communications among team members easier	2. How extensively people use it
3. Document communications	3. The intentions for which people use it
4. Make it possible for individual team members to expand their scope of influence on the project	4. The style and tone of communications
5. Make it possible for organizations to expand their scope of influence on the project	5. What people and organizations do with the information once they have it
6. Improve communications within the organizations overall	6. Whether a different communications style fits with the way the firm conducts its business

fortable using a keyboard to type. They prefer the freedom they feel when they can sketch and write freehand.

Some firms simply are not set up in a way that would make it easy to integrate groupware into their everyday business practices. People in these firms don't share information and don't need to know what each other are working on. Groupware would be irrelevant, perhaps even a distraction from an established, more individual focus.

3. *Intentions.* Groupware does not address in itself this area which, if left unattended, could cause significant problems on a project. Groupware is, after all, just a tool. It has the potential, if project team members intend, to help them substantially improve communications. If, however, their intention is more to cause problems, there is no filter in groupware to stop such misuse.

Whether project team members are cooperative or manipulative, helpful or antagonistic, working in the best interests of the project or in the best interests of their own paycheck, groupware places the same powerful tool for communicating in their hands.

Table 13.3   Possible Unintended Outcomes of Groupware

What groupware can do	Possible unintended outcomes
1. Improve 1-on-1 communications among team members	People use enhanced communications to exert individual influence at the expense of other players on the project team.
2. Make 1-on-1 communications among team members easier	Easier communications produces overload of information in the system. People have to take an inordinate amount of time to sift through the information they have.
3. Document communications	The knowledge that any communication in the system can be viewed by all team members inhibits people from bringing communication problems to light.
4. Make it possible for individual team members to expand their scope of influence on the project	People get involved in things that don't or should not concern them.
5. Make it possible for organizations to expand their scope of influence on the project	With more information at their disposal, people and organizations may become more argumentative, and, in the end, less effective communicators.

Much of the preceding text has focused on the potential of groupware to achieve gains in project communications. It is also important to explore some of the losses and problems it could create. Table 13.3 outlines these problems.

It is all too easy to envision how project team members could misuse groupware:

- Embarrassed by a mistake he has made, one subcontractor sends a memo to another, criticizing the quality of the other person's work. The memo is meant more to "cover" the first subcontractor's mistake and appears on the memo log for all project team members to read.

- Resisting suggestions to change a design, the project architect writes a memo to the contractor suggesting that the contractor is not being cooperative in working on resolving an ambiguous drawing.

- With the new information at her disposal, an engineer uses her expanded scope of influence on a project to showcase her own work at the expense of the overall project.

- With the overall level of communications information the groupware provides, the owner of the building becomes able to micromanage the project and attempts to hold people accountable for preexisting conditions they were unaware of.

- People more comfortable with using computers in general build up an edge in using groupware to exclude those who are less comfortable from discussions and to dominate the way the project takes shape.

Groupware itself works through its capabilities. Project team members can attempt to make rules and procedures for using groupware. However, the extent and intentions with which people use groupware are not built into its technology. The things that groupware does not necessarily influence are not, of course, the fault of groupware. Yet these areas can limit the extent to which groupware is used, and thus impact its effectiveness overall.

Using Partnering to Enhance Groupware Effectiveness

Groupware can enhance partnering significantly, but partnering is necessary to ensure that groupware is implemented fully and effectively. Groupware does not influence (or attempt to influence) precisely the aspects of communications that partnering takes on directly (see Table 13.4). Partnering methods have the potential to engage and improve the implementation of groupware, addressing items 1 and 2 on the right side of Table 13.2. In addition, it is possible to use partnering strategies to address how groupware is used, as noted in items 3, 4, 5, and 6.

Partnering methods could effectively address the problems of full use, fit, and intentions. At a preimplementation workshop,

Table 13.4 Using Partnering to Enhance Groupware Effectiveness

Groupware potentials	Groupware does not impact:	Groupware potential problems	How partnering enhances groupware
■ Improved every-day project communications ■ Improved individual participation in project communications ■ Improved organizational participation in project communications	■ How extensively or whether people use it ■ How people on the project team already conduct their business ■ The uses people make of groupware once they have it	*Full use:* The extent to which project team members use groupware *Fit:* The extent to which groupware fits with the way individuals and their firms do business *Intentions:* The style, tone, and intentions of communications sent through groupware	Project team members discuss how extensively they will use the groupware, raise issues, and resolve concerns for tone of use and determine specifics of use in ■ Goals statement ■ Communications procedures ■ Conflict resolution process

perhaps held as part of a preconstruction partnering workshop, participants would develop a goals statement, communications procedures, and conflict resolution process for using the groupware. They would monitor all of these in ongoing workshops over the course of the project.

The groupware goals statement addresses how project team members want to use groupware. If structured in the same way as the project goals statement, it will address full use, fit, and intentions with which people use groupware. Sample goals include the following:

- Use groupware for all communications. Project team members often benefit from knowing things that impact them indirectly.

- Groupware is for project work, not for criticizing people.

- Every project team member learns to type well enough to use groupware without going through a secretary.

Communications procedures will address in a very specific way how people will use groupware. Sample procedures include the following:

- People should use the informal memo form whenever possible.

- Meeting minutes will be prepared at the meeting on a laptop computer. All project team members will review the minutes before the meeting is over.

- Project meeting agendas will be posted in the groupware network at least one day before the meeting.

Groupware conflict resolution processes will spell out how to untangle some of the inevitable problems that will arise with groupware. Sample conflict resolution procedures include the following:

- Remember conflict is in itself not a negative thing.

- Try to work through conflicts rather than leave them in the background where they can affect project work.

- Try to resolve problems with the other person before bringing them to a supervisor.

- Meet with the other person face to face before going to a supervisor.

From these samples it is clear that partnering methods enable project team members to deal with problems of intentions and full use in working with groupware. Both kinds of problems are brought to everyone's attention, discussed, and resolved in some kind of common terminology in the goals statement and communications procedures.

The issue of fit is more difficult to work with, however, because it is more complex and at least partially outside the control of the people in the partnering session. People in a workshop can set goals and devise procedures for how to communicate and then take the responsibility for following through on their commitments in these areas.

Until people work with groupware for a while, however, it is difficult for them to predict how it will fit with their work. It is even more difficult for them to predict how it will fit with the way their firm conducts business. Setting goals for groupware use and discussing project communications procedures in detail help project team members to anticipate how groupware will fit in the way they and their firms do business.

In order to ensure that groupware is used fully and effectively, however, it is necessary that its use be discussed by all project team members at ongoing partnering workshops. This provides not only the opportunity to assess progress according to the goals the members have set, but also the ability to iron out problems of fit. Often, hearing how others are using groupware enables individuals to figure out how best to use groupware themselves.

First Person: Comments on Using Groupware from a Real Project Team

This sounds corny, I know, but I believe it's true: Groupware is a tool for the future. It's the basis for a whole new and much better way of doing business in design and construction.
Peter Johnson, Senior Vice President,
Spaulding and Slye, Developers

Of course I'm biased, I work here. But the truth is that Notes will save thousands and thousands of dollars on this project, reduce miscommunications and conflict, and improve the quality of the final product.

Bruce Sargent, Manager Of Corporate Real Estate,
Lotus Development Corporation

I am using it for the Rogers Street project but it is not my choice of communication for my other current projects. (This is mostly done by fax broadcasting of memos or handwritten notes—many times with sketches included.)

Linos Dounias, Principal, Arrow Street, Inc.,
Project Architect

The biggest thing for me is, groupware keeps me connected to the project. My contractual involvement on the project ended a month ago, but the groupware lets me keep up to date on what's going on. I can look at my computer and sort through project memos every day very easily and spot instances where our original design intentions might not be getting carried out. Then I can make the calls I need to keep our input intact.

Mary Killough, Principal, Stubbins Associates,
Interior Designer

Bibliography

David Kirkpatick, "Why Microsoft Can't Stop Lotus Notes," *Fortune,* December 12, 1994.

Elting Morison, *Men, Machines and Modern Times* (MIT Press, Cambridge, Mass., 1972).

Jeffrey F. Rayport and John J. Sviokla, "Managing in the Marketspace," *Harvard Business Review,* November-December, 1994.

Lee Sproul and Sara Kiesler, *Connections: New Ways Of Working in the Networked Organization* (MIT Press, Cambridge, Mass., 1991).

John R. Wilke, "Computer Links Erode Hierarchical Nature of Workplace Culture," *The Wall Street Journal,* Vol. CCXXII, No. 113, December 9, 1993.

4

Case Studies

This part synthesizes and applies the specific information in previous chapters into two case studies describing typical partnering applications: a comprehensive partnering effort and a "partnering-as-repair" effort for a project with "problems."

The writing in this section differs from that in the rest of the text. This part provides narratives, with dialog, characters, and even some plot. We take this approach to provide readers with more complete information and clear examples of how partnering works in the real world.

The situations in the cases are real in the sense that all of the project communications problems are issues we have encountered in many different partnering projects. The cases are fictional, however, in the sense that the characters and places do not exist. Any resemblance to real people and locations is purely unintentional.

14

Partnering by the Book

The Benson Management Services Building Case*

Comprehensive Partnering

What really happens to project communications when partnering goes "by the book," using all the available tools, skills, and strategies and starting before construction with the bidding process? This case examines such a complete example of partnering. The case explores a wide range of what partnering can do. When project communications improve, the impact for the design and construction process can be far-reaching. The case also illustrates in some detail how and why partnering works.

Characters

Jim Allen. Senior Architect working on the project

Charlie Anderson. Project Architect for the project

Tony Bruno. Wineski Site Construction Manager

*This chapter is based on a case study the author prepared for research conducted by the Space and Organization Research Group in the College of Architecture at the Massachusetts Institute of Technology; the research was funded by the Takenaka Corporation of Japan.

Ron Corman. Partnering Facilitator

Tom Craddock. Benson Facilities Manager

Joe Danehy. Benson Corporate Real Estate Vice President

Brian Murphy. Owner, Murphy Engineering (Electrical)

Robert Schmidt. On-site Architect's Representative

Tim Wells. Project Engineer, Air Engineering

Stan Wineski. Owner, Wineski Worldwide Contracting

Starting Point

It was the client, Joe Danehy, Benson Corporation's Vice President of Corporate Real Estate, who initiated partnering for Benson's new regional headquarters. It could have been the architect, contractor, developer, owner, or even the engineers, but in this case it was the corporate real estate department.

Joe Danehy took pride in being the pioneer; he liked to innovate. He recalls, "I kept hearing about partnering at seminars, conferences, and in the professional newsletters. At first I was not impressed. I've seen other programs come and go in the past, and in retrospect, they all seem like fads even though they sounded brilliant the first time I heard about them. Not so with partnering.

"I maintained my caution about partnering for some months, but the noise did not die down. Finally I had a chance to find out what it really was all about when I went to a national real estate conference. I ran into a few old buddies there who had used it on some recent projects.

"They went on and on about how it had saved them money and time, how it reduced conflict. What really hooked me on it, though, was when they said it actually made the job more enjoyable, more like it used to be in the old days.

"I made a mental note to try it the next chance I had, so when our new Mid-Central Regional Building came on-line I thought I'd give it a shot. I got all the booklets and brochures I could get my hands on from the AGC [Associated General Contractors] and AIA/ACEC [American Institute of Architects/American Consulting Engineers Council] and really tried to do it right.

"Starting right off the bat in the initial RFP's I let all bidders know [see Fig. 14.1, Exhibit A: Joe Danehy's initial letter] that this

---

Exhibit A: Joe Danehy's Initial Letter

December 1995

From: Joseph Danehy, Vice President, Real Estate

To: Bidders on Benson Contract 5483 Regional Office

Re: Partnering on the job

This brief memorandum is to inform all bidders on the new Benson Regional Office Facility Contract 5483 of Benson's intention to use Partnering methods and strategies in the design and construction of the new building. This will be Benson's first use of Partnering in a major facility, so it is important that it be done properly in order for the company to be able to evaluate the effectiveness of Partnering and its applicability to our specific needs.

Although we do not have direct experience with Partnering, we have researched the subject very thoroughly and believe that it has much to offer both us and our contractors. Because the new building is a research facility, it involves a particularly large number of potential communications problems Partnering can address:
- complex needs of the tenants
- government agency involvement
- strong personalities on the project team
- overall high costs to the company of any communications mistakes

If you win the contract, your requirement to Partnering is fourfold. You will:
1. Participate in several Partnering workshops before and during construction.
2. Make a good faith effort to implement decisions made in Partnering workshops.
3. Communicate the message and products of Partnering to all the participants on the project team.
4. Contribute to the direct costs of Partnering.

If you are selected to work on the job we will contact you regarding the selection of a Partnering facilitator. Meanwhile, please call me if you have any questions.

• Benson Management Services, Inc. • Corporate Real Estate •
• 1735 Lincoln Street • Chicago, Il 30374 • (906) 673-8000 • fax (906) 673-8100

Figure 14.1.   Exhibit A: Joe Danehy's initial letter.

would be a partnering project and that we expected that any successful bidder be prepared to participate fully.

"I didn't really have specific goals for partnering at the beginning. We hadn't had any huge problems on previous projects either in terms of dollars or in communications. I wanted to use it mostly out of curiosity, and I guess the other factor was enjoyment. Things in general work pretty well in this company. Like many of my peers, I'm always on the lookout for anything that could make some improvements.

"Maybe there was kind of a goal I did have, speaking of improvements. Just like everybody else these days, corporate real estate here at Benson is under pressure to demonstrate our worth. I guess I was looking to partnering to make a contribution to the bottom line, and to help us get more visibility with corporate. We thought we could use partnering to show our corporate-level managers how forward-thinking we are."

Initial Reactions and Starting Points

"I had no idea, when I started this thing, the way people would react to it," Joe Danehy reflects. "From where I sat it all seemed so obvious. I wasn't prepared for some of the comments I got in the beginning. I'm glad I stuck with it, some of the people almost made me drop it. They certainly got me to question what I thought I was doing."

Charlie Anderson, Project Architect, author of the winning proposal and principal at the design firm of Converse, Shapiro & Romanelli, didn't like the way partnering was presented to him. "It was all very ironic, very paradoxical," he recalls. "The client 'invites' you to participate in something that is supposed to improve commitment and collaboration. It seemed forced to me and I pushed back.

"In retrospect, I guess I overreacted. I think I reacted more strongly against the fact that we had to participate than against partnering in itself. I kind of liked the newness of partnering, I was very interested in the ideas."

Other team members had more trouble with the idea itself. Jim Allen, one of the other architects, saw partnering as a potential cost problem. "I looked at the letter explaining the workshops and all I could see was our billable time going up in smoke. It looked like a very bad idea for us."

Bob Schmidt, the architects' on-site rep, was also skeptical: "I didn't see the need for it. I looked at what partnering is supposed to do, and saw my job description. If I'm doing what I should be, keeping everyone informed, why should we need all these meetings?"

Mike Converse, the architecture firm's founding partner, also had reservations. "Sure I read some of the case studies of partnering and I was curious, but I can't say I really saw the need. I thought we already did most of what partnering was supposed to cover. If we didn't, why are we successful?"

Other members of the project team were more enthusiastic. Brian Murphy and Tim Wells, the electrical and HVAC engineers, looked forward to partnering. "People don't listen to us enough," Murphy explained. "There are so many projects where we see problems and bring them up but somehow the information gets lost. I couldn't be sure how partnering would work but it looked like it might get our input to be listened to a little more."

Tim Wells agreed, "We're the engineers, the underdogs. We don't get the glory or fame, but we can live with that. Not being heard is worse, though, and it happens too often. I looked at the partnering meetings and thought, 'Maybe they will help get our ideas onto the table a little more effectively.'"

Mike Sears, the facility manager Benson had slated to take the helm of the new building, was even more optimistic. "I saw the initial material on partnering and thought, finally, here's a chance for me to actually get some input into the building I will have to live with. All these other people are going to go away when the project is over but I'm not. It sure would be nice to have some say into how things are put together."

Stan Wineski, owner of the contracting firm that submitted the winning bid, and Tony Bruno, Wineski's on-site construction manager, were the most enthusiastic members of the project team on the partnering subject. "How could I be anything but enthusiastic?" notes Wineski. "We had already participated in a half dozen partnering efforts and they all helped in some way. Some were real knockouts, with big cost savings and time line reductions. Others were more basic, just keeping communications on an even keel. In every case, though, partnering was worth much more than the time and effort we put into it. I was delighted to see Joe Danehy bring this project in under the partnering umbrella."

"I love partnering," confesses Tony Bruno, Wineski's on-site manager. "I used to be a skeptic but my first experience with it changed all that, I could see a noticeable difference in how people got along. There was a lot less of the blame and finger-pointing we see so much of.

"I guess that out of everybody on the job, too, it might be that it really affects me the most. I'm the one who has to deal the most with people's communications, so I'm the one who sees all these problems up close. When people get along better, it makes my job a heck of a lot easier.

"I also get a kick out of the training. I never had any of that stuff in school, and I kind of like it. The facilitators have gotten us to do some pretty interesting things, building trust, improving listening skills, learning to run meetings more effectively."

The Facilitator Interviews

Ron Corman, the facilitator hired for the Benson project, encountered the project team members' perceptions of partnering when he interviewed them before the first workshop. While the wide differences of perceptions overwhelmed Joe Danehy, they were familiar to Ron from the numerous other jobs he had facilitated.

Ron conducted the interviews because he knew that the success of partnering depended in considerable part on the extent to which project team members supported it. Interviewing people was the first step in building project team member support to implement the commitments they make in partnering workshops.

Ron had discovered that he could effectively identify and address many of the concerns people had about partnering before any workshops occurred. With their concerns addressed, people could participate more fully in the workshop. Ron knew that the workshops represented a hefty sum of the participants' billable time, so he worked hard to make the workshop time as worthwhile as possible.

To select Ron, Joe Danehy had convened a subgroup including Charlie Anderson, Stan Wineski, and himself shortly after the winning bids were announced. The three of them went through a list of 11 local facilitators provided by the local Associated General Contractors' office. They each made phone calls to cover all 11 candidates and narrowed the group down to a short list of 3 finalists.

Joe, Charlie, and Stan split predictably on the three facilitator candidates. Stan wanted the retired construction manager because he had the most actual construction experience. "He knows the facts better than the others," Stan reasoned.

Charlie Anderson wanted the psychologist who had worked in several design firms. "We need somebody who understands people," he contended.

In the interest of satisfying the client, Charlie and Stan went along with Joe's choice of Ron Corman. Joe wanted Ron because of Ron's experience in many aspects of real estate, design and construction. He also favored Ron's custom approach. Joe thought Ron effectively balanced the flexibility of a custom approach with a core of definite building blocks in the partnering process.

From his perspective, Ron was pleased with Joe, Charlie, and Stan because they expressed an interest in taking a very active role in partnering. Ron knew this kind of active client participation is essential if the partnering effort is to be effective. Even though some clients were perfectly happy with the consultant-as-entertainer approach, it rankled Ron. "This is what gives our profession a bad name," Ron explained to Joe Danehy, "consultants and facilitators who go around dispensing 'wisdom' with no real connection to the project."

Ron's interviews turned up some useful information. He discovered, for example, that Charlie Anderson was willing to participate in partnering but worried that it might somehow compromise his design. Charlie made it clear that he had experimented with participative design back in the 1960s, that he was unhappy with the results, and that he didn't want to try that kind of thing again. Ron knew it would be important to differentiate partnering from such earlier endeavors.

Charlie explained his interest in staying involved with the design process in detail. "I learned my lesson after my first sizable building, an office building in the Southeast. It looked so good on paper, almost won an award in one of the glossies for 'Promising Young Designers.'

"I got busy while it was being built; turned my back on it while I did more design for other projects. When I finally got to visit it four days before the ribbon-cutting I couldn't believe what I saw. My name was on the sign outside but it was not the building I conceived. I could scarcely recognize it.

"That was when I made up my mind to somehow stay involved with any project I drew. I don't actually supervise construction, but I do try to stay involved enough so that as changes come up I can keep them true to my original intent."

Jim Allen's stance towards partnering was less dramatic than Charlie's and more like that of the rest of the participants. They were willing to attend the workshops and "see what happens."

Everyone also expressed hope that partnering would reduce conflict and litigation and improve communications and building quality. They thought that, on the face of it, partnering made sense. It made sense to have a preconstruction meeting. It made sense to have more frequent and better discussions about the project while it was being built. Everyone had an interest in making sure their concerns were represented in the project; everyone had some kind of story like Charlie's about how they learned to stay more involved in a project in order to see their own interests through.

However, everyone also expressed some concerns about the possibility that partnering would somehow weaken their ability to advocate for their own interests. No one wanted to land in court to resolve differences but at least everyone was familiar with how the courts worked. No one fully understood how partnering would unfold, so they were all reserving judgment—everyone except Stan Wineski, the contractor, and Tom Craddock, the Benson Facility Manager.

With Stan's experience with partnering on other projects, he had some clear understanding of how partnering would evolve on the project. He knew what it offered and he knew what he had to do to participate effectively. Tom did not know but he had goals and expectations that somehow partnering would increase his influence on the project.

Ron Corman used the information he got from the interviews in part to shape the agenda for the preconstruction workshop. He also tried to recall who said what so that he could refer back to the appropriate person as the workshop addressed relevant topics.

The Preconstruction Workshop

"We'll be working on two things today," Ron announced to the group. Joe Danehy, Jim Allen, Tom Craddock, the engineers, and all the subcontractors, a group totaling 21 sat comfortably, expectantly. Some listened attentively to Ron, some glanced through the thick program notebook, and some studied the room's interior fixtures.

Since all the participants arrived early, Ron started the meeting a few minutes ahead of schedule. Ron reviewed the agenda he had sent to the group, reading from notes he had made on a flip chart page for everyone to see. The group would work on seven items (see Fig. 14.2):

1. Partnering overview

2. Participants' interests and concerns

3. Goals statement

4. Communications procedures

5. Issue resolution

6. Communications skills training

7. Planning for follow through

1. Partnering Overview

For this brief segment, Ron did most of the talking. He explained partnering history, the different ways people used it on other pro-

---

1. Partnering overview
   - Partnering history
2. Participants' interests and concerns
3. Goals statement
4. Issue resolution process
5. Communications procedures
6. Introductory training
   - Applying the Myers–Briggs Type Indicator and building trust
   - Initial conflict resolution skills
7. Planning for the next meeting

Figure 14.2. Exhibit B: Benson preconstruction partnering agenda.

jects, and the range of successful experiences reported. He described the various components of partnering in detail. To help illustrate his points, he circulated publications on partnering from the AGC and the AIA/ACEC as well as the course outlines for seminars he taught.

Because he had done the interviews prior to the program, Ron knew Stan Wineski and Tony Bruno had participated in partnering before and were very satisfied with the results. He asked them to describe their experiences, responding to the group's several questions: Was it really worth the effort overall? Did it really make any difference in the end? Both Stan and Tony answered affirmatively, positively.

Though they had had a chance to ask him in the phone interviews and had not, several of the group members now asked Ron why the workshop was scheduled for only a day. Other partnering programs they had heard of lasted two or three days.

Ron responded that he thought initial workshops longer than a day usually resulted in an overload of ideas and intentions, and a gap between ideas and their successful implementation. Ron said he thought groups were more effective when they put their efforts into follow up and continuity.

2. Participants' Concerns

Here Ron broke the overall group into smaller groups of three and four participants, breaking up the cliques that had already started to form. He made sure to get people to sit with others they would have to communicate with once the project was up and running. Ron asked people to discuss and list both the hopes and concerns they had for partnering.

"Why are you doing this?" Tony Bruno asked Ron on the way to his newly assigned seat. "If you give people a chance to talk about their concerns, won't they just create issues and roadblocks that don't really exist? They didn't do it this way at the other program I attended."

"And what happened to the concerns people had?" asked Ron.

Tony paused. "I guess they didn't go away. People just talked about them during the breaks and lunches anyway."

"Exactly," Ron replied. "If people have concerns I'd rather know about them. That way at least I have a chance to respond to them.

Or more likely, I can get you or one of the others who has had experience with partnering to provide the responses. There's no reason people can't get their questions answered while they're here."

The concerns people brought up did not differ much from the ones they had raised in their phone conversations with Ron. The difference here was that they had the opportunity to talk them over with others and get their some bearings on where they stood.

Ron particularly enjoyed the "What if...?" questions the group raised. "What if we get to a point where we just can't settle a disagreement among ourselves?" "What if we reach all these agreements and somebody doesn't pull his or her own weight?" "What if somebody lies?" "What if people don't change?"

Ron reminded the group that partnering was not intended to alter human nature, just to make it easier to get along. On the other hand, the program ought to improve the odds for effective communications and improve quality as a result.

3. The Goals Statement

The first time Ron worked with a partnering group to produce a goals statement, he was skeptical about achieving results. It seemed artificial to him: get people to brainstorm a list of idealistic goals, sign the list, and keep it around the project. Ron's own track record with New Year's resolutions was not great. What would get the group focused to implement their goals?

Ron also knew from basic organization development experience that goals statements could be useful if groups put them together effectively and actively and frequently used them to monitor progress. He also knew from speaking with people who had participated in partnering that the goals statement was widely used and that people knew it could produce results.

The first few times Ron worked with groups to produce a goals statement, he was most impressed by their energy. People enjoyed working on the documents; they became exhilarated talking about possibilities and hopes. When Ron mentioned the level of energy he noticed to several of the group members, they explained, "We spend so much time and effort working out problems, we seldom get the opportunity to talk about anything positive."

Ron also noted that people who might tend to argue with each other as the project unfolded would probably find it harder to be unreasonable with someone they had worked with on defining shared goals.

Working on the goals statement enabled people to see, early on, that they agreed on a great deal. Thus working on the statement provided an early success for the group, a success they would need when they addressed more difficult issues later on.

The Benson project team took about an hour and a half to come up with their goals statement. Each subgroup produced four or five goals, which Ron listed on the flip chart. Once they were finished, it was easy to combine duplicates, tune up the wording, and produce a working list overall (see Fig. 14.3).

These are the partnering goals formulated and agreed to by the partnering group:

- The building should come in on schedule, under budget, and with a profit.
- All communications should show respect for all involved.
- There should be timely solutions to problems and conflicts.
- We should work at improved coordination and better understanding of other trades and activities during construction.
- There should be minimum disruption to the neighborhood.
- We all should deal *directly* with each other when there are conflicts.
- Quality building.
- Project will go smoothly, be profitable, and fun.
- Understand others' points of view.
- Less conflict.
- Help maintain building design intent.
- Architect will turn around RFIs in 24 hours or less.
- Resolve conflicts quickly.
- Monitor and measure partnering mechanisms.

Figure 14.3.   Benson Office Building partnering goals.

Ron ended the goals statement exercise with the action he had found helped make the statement effective as a project tool: he asked all the people in the room to sign it.

At first several people hesitated. "You really mean this, don't you?" Charlie Anderson asked. It was Charlie who had specified the goal about preserving the building's original design intent in the course of making any changes. The contractors, Stan and Tony, were not very enthusiastic about that goal but they agreed to it after Charlie explained what he meant and why it was important to him.

Ron asked the group to think of ways to bring the goals statement to life while they took a break. "Now that we have it in writing, we have to work on making it a working document," he advised.

Implementing Goals.   Led by Stan Wineski, the group returned from break quickly. "We did a few great things on other projects that maybe we could learn from on this one when it comes to implementing the goals statement," he addressed the group. "We had all the firms put the statement in peoples' pay envelopes on a regular basis. Then, we put it on the agenda for every project meeting, just to keep in the background. The other thing we did, and you can laugh at this, but it worked, we put it on T-shirts. By the time we were a few months into the project, you could bet on seeing at least one person wearing the thing every day."

Before the group could either reject or approve Stan's ideas, Ron asked for two volunteers to work with Stan to develop recommendations for implementing the goals statement to bring back to the group at the follow-up workshop. Charlie Anderson and Joe Danehy raised their hands quickly, so Ron moved on to the communications procedures.

4. Communications Procedures

The architect Jim Allen provided the chief point of discussion to begin the work on communications procedures. "We won't have many meetings here, will we?" he asked. "They're such a waste of time, and you all usually like to spend your time on the job instead, don't you? I was thinking we'd do the standard monthly show-and-tell. Beyond that, we can always meet as needed, right?"

Stan Wineski and Tony Bruno argued on behalf of more frequent communications. "This is another thing we learned in past partnering," Stan pointed out. "We used to approach meetings the same way as you, Jim, but we've changed. We used to have all kinds of communications problems, mostly about key people not knowing key information about the project.

"In a partnering program over a year ago, the facilitator got us going on meeting weekly, not one of those show-and-tell things, but a meeting where everyone gets the information he or she needs. He showed us how to run meetings so that everybody talks; there would be no wallflowers who hang back and then complain later.

"At first we couldn't see the point. But after we did it a few weeks we could see that the hour spent in the meeting was saving us many more hours during the week when we would normally be chasing each other down. We also found we were doing a lot less rework, having a lot less frustration because we had the information we needed."

"It's a good thing we got used to that one a year ago," Tony Bruno added, "because at a program we did a few months ago we started having daily meetings."

Jim Allen was incredulous. "You can't be serious," he protested. "You want our site representative to be there every day for a daily meeting? What do you do there? Make a quilt? I think you may be taking this whole partnering thing just a little too far."

"We thought the same thing," Stan Wineski stepped in. "But we tried it and I can tell you it saves a great deal of time. How much time do you figure gets wasted on the average job with people just trying to find each other? You know how easy it is to get 'lost' on a big project.

"We figured that people drink coffee anyway—they might as well drink it in a meeting. We meet every day when the coffee truck comes in. We get our coffee and come back to the construction trailer. We don't have chairs, we stand up and do a quick go-round through all the subcontractors and major players. The client can be there, in any case he knows where and how to find us if he needs to.

"It's been great. We all get better information, and the meeting only takes about 15 minutes."

"Let me add something to the good ideas you already have," Ron Corman suggested. "How do you keep track of what's happened? Do you document the meeting in any way?"

"That's one thing we haven't figured out," Stan admitted. "We tried taking minutes and that took too much time. Then we just let it go and that hasn't been good because you need to remember what we agreed on. What's your idea?"

"In a few of the jobs I worked on, they kept a flip chart in the trailer, set up like this." Ron took a flip chart page and drew a line down the middle. On the left side he wrote "Issues." On the right side he wrote "Actions."

"Brilliant," Tony reflected. "I can see it. People come into the trailer and start listing items even before the meeting starts. Then during the meeting, the list keeps people focused. As you write down the actions, people get to see if what's being written is what they thought they agreed to. Then you just leave the whole thing in the trailer and you've got your documentation of who is supposed to do what."

"Exactly," smiled Ron, and began to get the group to write up what they had been discussing. He was pleased with the results (see Fig. 14.4, communications procedures), particularly the two new ideas that this group came up with, innovations that Ron had not encountered before.

First, they came up with the idea to exchange home phone numbers among all key decision makers. The second innovation this group came up with was the idea of several scheduled weekly conference calls among all the major players.

5. Issue Resolution Mechanisms

Though it was time for lunch, the group asked Ron if they could work through the meal. Stan pointed out, "We're used to getting up at 5:00 A.M. This may be second breakfast time to you but to us it's almost dinner."

Ron agreed and got the group started on issue resolution while they settled in with sandwiches brought in from the hotel. Ron thought issue resolution made a nice counterpoint to the goals statement and communications procedures, as the issue resolu-

To maximize communications effectiveness on the project, we will use the following procedures:

1. Weekly job meetings, all subs and major contractors. Leadership rotated among all participants.

   - Review project status, each participant reports.
   - Identify, discuss, and resolve problems. Take a proactive approach.
   - Anticipate and discuss short- and long-term issues.
   - Open forum question-and-answer session.

2. Daily on-site meetings.

   - 10 to 20 minutes
   - Tony runs meetings, some rep attends from each sub- and major contractor.
   - At coffee time, 9 A.M.
   - Accumulating issues agenda; use white board.

3. Several times per week, conference call of key players.

   - Charlie, Mike, and Stan to coordinate.

4. Emergency contact prearranged.

   - Phone access provided to all key players.

Figure 13.4.   Exhibit D: Benson project communications procedures.

tion process focused more on problems than on interests or methods. Issue resolution defines and specifies steps people will follow when conflicts arise.

"Do we really have to do this issue stuff?" asked Brian Murphy, the electrical contractor for the project. "We've spent so much time spelling out goals and procedures, shouldn't that do it? If people just follow the goals and procedures we wrote, everything should be all right. We did all sign them. It's not like anyone is going to break the promises they've made."

Stan Wineski and Tony Bruno responded. "No way," noted Stan. "Even with the best goals and procedures, people will miscommunicate. You've got to have these procedures spelled out ahead of time. I couldn't believe the grief and aggravation that it saved on the last job. Because people knew where to go, who to talk to when they had a problem, it really made things easier."

"He's absolutely right," Tony added. "I had this one argument with the plumbing people. They wanted to put their installations in ahead of us, and they just wouldn't listen to our side.

"We went back and looked at our issue resolution mechanisms and found the statement we had written that said if there was a conflict between any of the project team members and they couldn't work it out themselves, they would involve two other unbiased team members to help them. We went to the architect's on-site man and the HVAC contractor, two people we both respected. They sat with us for an hour or so and they did it, they helped us map out a schedule that worked for both of us."

"I like this," Joe Danehy reflected, listening to Tony and Stan's points. "This is great because it gets us to lay out procedures for conflict resolution now when we're all in a positive frame of mind. I also like it because it's realistic. It says we will fight, let's be realistic about that.

"We've all had countless problems before on projects, yet we sit at the beginning of a new project and it's as if we never had an argument. Instead of trying to make believe conflict won't happen, let's plan for it thoughtfully and let's figure out how to manage it when it comes up."

Tim Wells now addressed the whole group for the first time in the workshop since he introduced himself in the initial go-round. "I'm especially looking forward to this," he observed, "because it may actually get me into the loop. So many times on a project I don't get enough information about what's going on. Then when I do find out it's kind of late, so when I raise my point of view it causes conflict. It's usually better, at least easier if I just keep my mouth shut, but then the project suffers and I don't like to see that. If we had clearer procedures about how to resolve differences of opinion, we'd all be more likely to bring out information everyone needs to know."

Ron put the group to work then, in much the same way as they worked on the goals statement, in small groups of three and four participants. He made sure everyone was sitting in new groups, and he tried to get people to sit together in these groups who would later likely be in some sort of conflict with one another. He was pleased with the final document they produced (see Fig. 14.5, issue resolution mechanisms).

6. Training

"Can we rest now?" Jim Allen asked, finishing a second piece of chocolate cake to top off his lunch. "We're not going to take this training topic seriously are we? I mean, we all know how to communicate. If we want to learn that kind of thing, we could take a seminar, couldn't we?"

Jim looked at Stan and Tony, expecting them to disagree. "Okay, okay, don't say it, I know. You did this in your last partnering workshop and it changed your life, right?"

"Close," replied Stan, smiling. "Come on, Jim, you know you stand to learn some useful things here, like listening skills. For myself, I find this part of a workshop is the part I usually get something out of personally, not just for the sake of the project. It's the training I can usually apply to other areas of my life.

"I like the training, too," Tony added, "but the real point of it is that it's the thing that makes all the documents work. There's nothing like putting together a high-minded goals statement and then undermining it the next day when we get into a disagreement on the job."

With the few hours of time remaining, Ron Corman briefly introduced the group to two subjects: the Myers–Briggs Type Indicator, a research-based personality profile, and basic skills for self-mediation and conflict resolution. He knew the group would not have enough time to become skilled in either area but he believed they could learn enough to carry several core ideas back to the job.

Ron was pleased that Stan Wineski had spoken on behalf of the need for training. This was the part of partnering that Ron found the most frustrating: the fact that many people didn't know how much they didn't know when it came to communications skills.

a. Myers–Briggs Type Indicator.   Ron had a fair amount of success in teaching mediation skills with a two-part approach. He began not with the skills themselves but with the Myers–Briggs

We the project team agree to follow the following procedures to resolve conflicts if they arise.

1. In all matters, we will attempt to resolve conflicts quickly, efficiently and directly, and at whatever level on the project where they arise.

2. People will take responsibility for trying to resolve their conflicts.

3. In any conflict resolution discussion, both people will try to work on solving the problem in a way that benefits both people, rather than just on ways that win primarily at the other person's expense.

   ■ In discussions, both parties will address each other with respect.

4. If people disagree, the first step they will take is to meet on their own for at least an hour, trying to work out a conclusion satisfactory to all.

5. If the first attempt at conflict resolution fails, they will try again within a week.

6. If the second attempt at resolving the conflict fails, they will involve their supervisors.

7. Conflict resolution discussions will take place at a location agreeable to both parties.

8. People will try to resolve conflicts directly with each other but also work within chain of command in the organizations. The chain of command for resolving conflicts is

   ■ For Wineski, first Tony then Stan.
   ■ For Stillman & Converse, first site representative, then project architect, then principal in charge.
   ■ For subcontractors, general contractor.

9. Verbal discussions are to be followed with a letter, but start with discussions.

Figure 14.5. Exhibit E: Issue resolution mechanisms.

Type Indicator (MBTI). The MBTI is a research-based personality profile that characterizes peoples' preferences in communications, decision making, and use of information. The MBTI is one of about 20 different personality profiles on the market and available to facilitators. Ron preferred it over the others because it provided the most insight in the shortest period of time. (See Chap. 10, titled Valuing Differences, for more detail on the MBTI.)

Ron was not a psychologist but he was comfortable and skilled with the MBTI, having administered a shortened version of it to several thousand people in the eight years he had used it. Ron liked the MBTI because he had seen group members use it to improve their ability to understand each other, resolve conflicts, and identify and address shared problems. Often, people who had worked together for years would come out of a few hours' work with the MBTI and remark to each other, "I finally understand you!"

Best of all, Ron had seen the MBTI get people to be able to laugh at themselves and take an honest look at themselves and the impact they had on the group. Ron liked to think of it as a "mirror of character."

Ron was not surprised to see that in the Benson project team most of the engineers came out with (there are data on the personality types of different professions) personality types like engineers (detail-oriented, logical, sequential thinkers) and most of the architects came up with types like most architects (more intuitive, visual, abstract, less detail-oriented, and somewhat more feeling). He was very surprised (though he shouldn't have been, based on Stan's participation) that Stan Wineski came up with the personality type most common to ministers rather than the one most common to general contractors.

Ron asked the members of the group what the MBTI explained to them. That, he knew, was the key question. People liked to read about and discuss their types, but it was seeing the instrument in action that made it useful. Some of the things the group said the MBTI explained to them were:

- Why architects don't provide enough detail for engineers and contractors.

- Why engineers and contractors provide more detail than architects want.

- Why the motto of the first design firm Charlie Anderson worked in was "Keep It Fuzzy" while the motto of Tim Wells' first engineering firm was "For want of a nail...."
- Why the motto of Jim Allen's alma mater design school was "All Great Architecture Leaks."
- Why Joe Danehy was so compulsive about packing his trunk.
- Why Tony Bruno was such a stickler about punctuality (he and Joe Danehy were the only two members of the group who passed Ron Corman's "High J" test of time focus: they both had clocks in their bathroom, and all the (many) clocks in their house were carefully synchronized.
- Why Jim Allen's car was a mess and it didn't bother him. Why he usually spilled his lunch and had a hard time making a bed.

Brian Murphy expanded, "I ran into the MBTI before in counseling the church had my wife and I do before we got married. I found that it really does help us to understand each other. We find that a lot of the arguments we might have we're able to settle because what we're really arguing about is different preferences we have for getting a job done.

"It's like that on the job. If it helps us understand each other better, it should help us communicate."

Ron explained that they could spend many more hours on the MBTI, which would have suited several of them just fine. Mike Sears in particular was interested in how the MBTI might help him manage his time more effectively. Several of the men said they wanted to bring it home to give to their spouses and children.

b. Conflict Resolution Skills.   Ron Corman switched gears to introduce conflict resolution skills. He brought out the communications procedures sheet the group had produced earlier and taped it to the wall. Then he asked the group to list the inevitable arguments people knew they would have with each other.

People hesitated at first but then started to write. After a few minutes, everyone was looking up at Ron, their lists completed. Ron worked his way through the group, listing the situations on a fresh sheet of flip chart paper:

- Tony Bruno anticipated he would argue with Jim Allen about slow turnaround for RFIs (requests for information).

- Charlie Anderson thought he would argue with Stan Wineski about the changes and substitutions of materials he had specified.

- Mike Sears thought he would probably argue most with his own boss, Joe Danehy. Mike thought he would be wanting to make changes to make the building more maintainable and durable where Joe would be focusing more on external appearances and on cost savings.

- Brian Murphy thought he would argue with many of the other subcontractors about scheduling priorities.

- Joe Danehy thought he would disagree with everyone about making changes as the building was going up.

Ron listed a few other situations on the flip chart and then called people from the group to role-play with him how the situations might go. After identifying where the destructive conversational patterns came up in recent jobs, the groups moved on to working with skills. For this, Ron brought pairs of people up in front of the group to work in a real situation they thought they might encounter.

Charlie Anderson and Stan Wineski came up to discuss substitutions. Brian Murphy and Tim Wells argued about scheduling priorities. Ron had them stop at the points where the conversations turned unproductive. He then asked them to try new skills for listening and giving information descriptively to work in a collaborative way on the problems.

The group struggled with the skills at first but then quickly developed a basic ability to use them. "I wish we had a chance to do this on our last job," Charlie Anderson reflected. It could have saved a lot of arguing."

7. Planning for Follow-Up Activities

Ron Corman began, "We need to schedule a follow-up workshop now if we're going to make all our good work for today real and bring it back into the project. When would be a good time to do that?" After about five minutes of discussion, the group settled on a date about two months into the future.

"That's it, ladies and gentlemen," Ron concluded, asking them to complete and leave their evaluation or feedback forms on the table at the door on their way out. The forms came back as they usually did at the end of an initial workshop.

"Interesting, I learned a lot."—Charlie Anderson

"I liked the personality profile."—Jim Wells

"I liked the program but thought we spent too much time on the personality profile."—Jim Allen

"Can't evaluate until the building is completed and I'm in."—Mike Sears

Follow-Up Workshop 1: Interviews

Ron Corman sent all the members of the project team copies of the goals statement, issue resolution mechanism, and communications procedures they had produced at the initial workshop about a week after the workshop was run. He kept in contact with the project after the workshop with a weekly phone call to some member of the project team. Ron was pleased to find out that the team had tried both the weekly and daily meetings and that they were going well.

"I can't believe I'm saying this," Jim Allen reported, but it's more than worth our time to have Bob Schmidt attend." Joe Danehy also thought the scheduled meetings were proving to be very useful.

As the May follow-up program approached, though, Ron was struck by the team members' reluctance to raise issues they were encountering on the project. "Everything's going great, just great," Charlie Anderson reported, about three weeks after the initial workshop. "We have no problems, nothing to speak of. We're all very busy and happy to be at work on the project after all the delays caused by the horrible winter we've just had.

"Are you sure we really need the follow-up, Ron? I've been talking to a number of the others, at those daily meetings. They asked me to ask you, can't we put it off?"

"I don't know," Ron replied. "If you really don't need it, I suppose you shouldn't have to have it. On the other hand, it is in the contract and unless you're some kind of magician, you will probably be running into some communications problems by the time the meeting date comes around. Are you sure there are really no communications issues on the job right now?"

Charlie was silent for a few seconds, just a bit longer than his usual pause before answering. "Well, there's this little thing...."

"This little thing," Ron reflected, that was the way the others had put it. There were no real problems, they didn't really need to meet. On the other hand, there was "this little thing." Each of them had some "little thing" they hadn't quite gotten around to resolving, just a small problem.

After he completed calling all the members of the project team, Ron's list of the "little things" included the following:

- Mike Sears was very upset at the number of cement trucks coming onto the site, the route they had chosen to take through the neighborhood, and the early hour most of them came charging through.

- Tony Bruno was already several weeks behind schedule because the architecture firm had not returned several of his requests for information. In one case, the material Charlie Anderson had specified was no longer in production. Tony needed Charlie to recommend a substitute, which he had done by phone but not backed up with the necessary paperwork.

- Tim Wells and Brian Murphy had gotten into a major conflict about scheduling priorities. Once Tim found out that Tony could not get the materials he needed, he argued that he should go ahead and do the HVAC ducting in the basement before Brian Murphy got in with the electrical; he contended it would give a better job in the end. Brian, meanwhile, had already planned a ski vacation around the scheduled project phasing, and was not pleased about having to make changes.

- Everyone complained that Mike Sears was "too pushy." Mike, on the other hand, complained that people were avoiding him.

Ron kept this list in his notes. He would not use it to influence the program, he believed the group benefitted from creating its own agenda in the meeting itself. On the other hand, if the group did not include these issues, he would be sure to steer it to make certain to cover the issues.

Follow-Up Workshop 1: Program

Two months can be a long time in the midwest when it comes to weather. It was hard to believe on the mid-May day of the follow-up workshop that there had ever been snow on the ground. It was past spring, right into summer weather, with the temperature in the low 80s and the humidity making things uncomfortable.

Ron noticed immediately that the group was not in a positive frame of mind. The mood in the room felt very different than it had been two months earlier. Where people had been upbeat, optimistic, and gregarious, they were now quiet, sullen, and even openly disagreeable.

"I'm not sure if anyone has told you this," Bob Schmidt (the architect's on-site representative) confided to Ron as they both poured coffee, "but this is really not a good time for this. A few of us feel like this is not a good use of our time. We should be out on the job. If it wasn't for Mr. Danehy pushing for this meeting, that's where we would be. I don't see how he expects us to get the work done if we're not on the job." Ron replied that it appeared to him that there were definitely a few items to work on and that he would try to make the day productive for everyone.

Ron began by breaking the group into subgroups and asking each to make up a flip chart page listing things that were going well (indicated by a plus sign) and things that were not (indicated by a minus sign) on the project at the moment. The groups took about a half hour to come up with lists that included all the four issues Ron Corman wanted to cover along with a few others that also seemed important (see Table 14.1, project status at follow-up workshop 1).

Before he put the group to work on the issues on the list, Ron asked the group how they felt about the project now, and how they felt about partnering. Their response to both questions was the same: "Disappointed."

Charlie Anderson volunteered, "Hate to be the one to say, 'I told you so,' but some of us feel that the last workshop really didn't accomplish much at all. It's too bad. For all my complaining, I really thought this one would be different. I was so positive after our last workshop. I can't see how it could have gone wrong."

Table 14.1 Exhibit E: Project Status at Follow-Up Workshop 1

Positive issues (+)	Negative issues (−)
▪ No complaint calls (Stan W.)	▪ Architect and engineer slow to respond to contractors on drawings and change orders
▪ Contractor staff easy to work with	▪ Lapses in communication between major team players
▪ High level of attention to quality	▪ Client representative unreasonable and inflexible
▪ Good cooperation among most subcontractors	▪ Short-term schedules not accurate enough for client
▪ Low level of friction and conflict on the site	▪ Lapses in advance planning; too much firefighting
▪ Attempt at fairness in most conversations	▪ Low profit
▪ Partnering has helped	▪ Meetings are OK but could be more participative
▪ People have tried to make improvements	▪ Construction seems slow

"Me too," Jim Allen agreed. "What about you, 'Mr. Partnering'?" Jim referred to Stan Wineski.

"How about 'Stop whining and let's get to work,'" Stan replied. "Remember we said at the first workshop that conflict is inevitable. I really don't see the disagreements we're having as much of a problem. They will be if we don't work on them, of course, but we're going to work on them now."

Ron Corman shook his head in agreement, reflecting on the point Stan had made. It was a phenomenon he saw all the time, that people who worked at partnering felt they failed if a difference of opinion came up after a workshop. Ron kept trying to tell people, partnering does not eliminate differences of opinion, it just provides tools to manage them wisely.

Then Ron began with problem solving, one of his favorite partnering activities. He broke the group into subgroups and assigned each to one of the tasks raised by the overall group: architect and/or engineer response to requests, lapses in communication, planning, and short-term schedules. Ron decided to save the Mike Sears issue a bit until the group got some practice. Ron then told each group it would be responsible for coming up with some ideas for addressing the problem.

When people brought problems to the group, they usually presented them in a tone that implicitly asked, "What are you going to do about my problem?" In partnering, the person who experiences a problem sits down and works with the person who is creating it; they try to come up with something that works for both of them. Ron thought this was the real genius of partnering, to get people to work with each other actively to resolve their mutual problems.

With the complaints about the architecture and engineering firm's slow turnaround of shop drawings, change orders, and RFIs, Ron had several of the people from the architecture firm sit with several of the people from the contracting firm, the client, and several key subcontractors. Together they had to define the problems and begin to take some steps to resolve them. At first they looked surprised to be sitting at the table working with each other.

Ron responded, "It's the facilitator's job to get the problem solved but not to do the actual solving. I could tell you 20 ways to solve the problem you're working on, but none of them would work as well as the ways you come up with yourselves. After all, the solutions have to work for you.

The groups worked for about a half hour; then Ron provided some training on group problem solving. Using a very basic three-step approach, he had the groups first list "aspects of the problem." He stressed that they may not come up with any solutions right away, but should begin instead by fully chronicling the dimensions of the problem. Ron knew that teams have little trouble coming up with quick fixes, but they may result in an incomplete or misguided solution to a problem. He also knew this step helped groups see the complexity of many of the problems they face.

Once the group listed 10 to 15 aspects of the problem, Ron had them list possible action steps, focusing for the moment on coming up with many alternatives rather than instantly trying to force the "correct" answer. Finally, Ron had the groups look at the alternatives they had listed and choose several to begin to work with. Throughout all three steps, Ron stressed to the groups to document all ideas on the flip chart so that everyone could work with the group's accumulated insights from the same page.

As the subgroup discussions were going well, he let the participants extend their time a bit. Then he asked them to present what they had come up with for the overall group to discuss, review, and ultimately agree to. Ron let the groups work for about half an hour, then they were to present their ideas to the whole group for discussion, refinement, and feedback. All four groups came up with practical plans for addressing the issues they worked on. None of the ideas was terribly creative or innovative, but each was effective in breaking up the logjam of communications in the area that had the problem.

In the case of the slow RFI turnaround, the group came up with several strategies (see Table 14.2 regarding problem solving at the follow-up workshop):

- Contractors would prioritize their requests so the architect and engineer would know what was really important.
- Contractors would try to resolve questions verbally without always using the bureaucracy of formal requests.
- The architecture and engineering firm would look into its internal processes to determine ways to speed up the turnaround.
- One person in the architecture and engineering (A/E) firm would be designated to keep a log of all communications. The log would be updated weekly and distributed. It would be available on request from any project team member.

The other groups came up with similar solutions to the problems they addressed. All the groups agreed to provide a brief status report on their activities in the weekly project meetings. When they were finished, Ron passed the marker so people could sign the agreed upon plans much as they had done with the goals statement.

Though he was not comfortable with it, Ron knew he had to address the Mike Sears issue next. "How do you want to handle this, Mike?" he asked. "The group wants to discuss how to work with you.

"Would you like to stay in the room here or would you prefer to get some air while we try to clarify what the issues are? If it's pos-

Table 14.2 Exhibit G: Problem Solving at the Follow-Up Workshop

Aspects of the problem	Possible actions	Things to try
<ul><li>RFIs are slow in coming back to contractor</li><li>When RFIs do come back they are incomplete and often wrong</li><li>Contractor is delayed</li><li>RFIs are not completed in sufficient detail so the A/E firm can understand them</li><li>Different subcontractors complete RFIs in different details and languages</li><li>Contractors can't find where RFIs are in the pipeline in A/E firm</li></ul>	<ul><li>Handle everything with RFIs</li><li>Handle nothing with RFIs</li><li>Have someone from the contracting firm work in the A/E firm</li><li>Contractors mellow out</li><li>A/E firm take life more seriously</li><li>Both sides treat each other with more respect</li></ul>	<ul><li>Contractors prioritize requests</li><li>Contractors try to settle everything by phone first before going to a written RFI</li><li>A/E firm to look into internal communications gaps and report back to project team in one week</li><li>One person in the A/E firm is designated to maintain RFI log</li><li>A/E firm representatives on project site one day each week to draw out responses; try to settle problems less on their own after returning to office</li></ul>

sible I think it's better if you stay, but I know sometimes that's just too uncomfortable to work."

"I'll stay," Mike agreed.

"Great," Ron responded, and broke the group up into subgroups once again. Once again he used the three-step creative problem-solving process. Once again the groups came up with specific action steps. Mike himself moved from one group to the next, sometimes listening and sometimes providing information, trying not to be defensive.

Although the discussion was a little more personal and volatile than the earlier topics, the problem unfolded in much the same way. Initially it looked like a simple issue: the client's facilities manager was being unreasonable, making demands, making changes, not providing adequate lead time, and not being willing to respond to input from the other members of the project team.

With more detailed discussion, however, it turned out that the underlying problem had less to do with Mike and more to do with Mike's relationship with Joe Danehy. Joe had told Mike several times that he was counting on him to make the building something special for Benson, that it was to be a key building. He had also instructed Mike not to "take anything" from the other members of the project team. All Mike really had done was to listen to his boss, who was inadvertently undermining the very partnering process he was sponsoring.

As Joe Danehy listened to the discussions he began to realize the tangle he had created. He volunteered to have lunch with Mike and two of the people who had complained the most about him.

With the more difficult work of the day completed, Ron asked the group to think about the goals statement. He took out the original piece of flip chart paper, now two months old, and taped it on the wall while the group ate lunch.

After lunch, Ron asked the group to rate how they thought the goals were holding up. After some discussion, people gravitated to the mention of profit as a problem. Things were going reasonably well; nearly everyone had worked on projects that were much worse. If there was anything to improve, everyone agreed, it would be to find some ways to actually make a profit on the project. Everyone was concerned that Benson's original low margins would quickly evaporate and that they would be left with a project that was interesting professionally but did little or anything for their organization's revenues.

Ron organized small groups once again. He cautioned the groups to move slowly and worked with them to use several techniques for visualizing and imaging. He gave the groups creative materials to work with and showed them how and why to integrate play into their work. After a bit more than an hour and a half, one of the groups announced that it had devised a tentative approach for cutting the schedule. It wouldn't add up to a large number of days, but it was possible that enough days could be involved so that it would improve everyone's profit.

The other groups worked with the idea for the remainder of the afternoon. It was difficult to get people to do something they usually don't do, reviewing the work of people outside their own organization and field to trim the schedule.

Joe Danehy was amazed at the end of the exercise. "Every time I brought up the subject of cutting the schedule, people looked at me as if I had asked them for a gift. Yet today, they all worked together at it and they came up with what looks like some significant gains."

Additional Workshops

Ron Corman continued the process of calling project team members between workshops and planning the agendas for two more sessions. He ran the workshops as he had the first follow-up meeting. Half of the time was devoted to solving immediate, specific problems. The other half was given over to tackling larger issues that faced the whole group. The programs continued to be effective and actually increased in effectiveness as people learned what to expect and how to use the time and structure of partnering most productively.

Still, Ron was not surprised when Joe Danehy terminated the contract for partnering going into the second year. Often, too often, partnering clients canceled a contract once the immediate problems were solved and the groundwork for communications was in place.

"You've done a great job, Ron," Joe reassured Corman. "We all appreciate what you've done for the project and what you've done for us individually. We're just starting to feel that we could use the partnering time better on the job. There haven't been any real problems in awhile, maybe because you were so effective at the outset. So what we'd like to do is put the second year sessions on hold and see if a need comes up."

Ron tried to argue, pointing out that new problems would be coming up in the later phases of construction, but Joe was inflexible. "Budget cuts," Joe explained, "they get us where it hurts."

Ron knew from past experience that it would probably not be worth his time to argue further, as Joe's decision was made and signed off on by his boss and his boss' boss. Still he felt he had to say that he thought it was a mistake to take away the forum partnering had provided. He thought the project would benefit from it, he knew it was worth the time in pure project cost terms.

Ron was not surprised four months later either, when Joe Danehy called and asked him to take on a mediation case between Stan Wineski and Charlie Anderson—nothing major, "just a little thing."

Commentary: Observations

This case illustrates several recurring themes that occur in real partnering projects:

- People underestimate the need to follow through, especially after a successful initial workshop.
- Different people come into partnering with very different levels of experience and expectations.
- When some project team members have participated in other partnering workshops, it can help provide a baseline and benchmark for other group members.
- The same group members who have participated in other workshops may assume that the way they learned is "the" way to do partnering.
- The initial workshop headed off many potential conflicts and miscommunications.
- The work with personality profiles accelerated the project team members' getting to know each other.
- The communications skills training both provided skills and anticipated specific problems.
- Partnering raises internal issues for the organizations that participate.
- The mood and focus differ greatly between initial workshops and follow-up workshops.
- Follow-up workshops provided a needed forum to address project issues.

- In follow-up workshops, the groups did have the ability to solve many of the problems they brought in. In some cases, they actually made it look easy.
- The three partnering documents work together to achieve gains in communications.
- Follow-up workshops make people accountable for their actions by providing a forum in which all kinds of project issues get discussed.
- Although follow through may have been a problem, boredom and apathy were never issues in the workshop.

Commentary: Improving Partnering Effectiveness

This case illustrates factors that help improve the effectiveness of the partnering effort.

- The interviews the facilitator conducts before the workshop help plan the program and make it more relevant to peoples' concerns.
- It is important to take the time to get the right people to attend.
- In the workshop it is important to get full participation—from all participants.
- Breaking the large group into subgroups helps achieve full participation.
- It is important to have people sign the documents.
- Much of the success of the original workshop can be attributed to the project team members' ability to implement the results on the job.
- The facilitator has three major tasks to perform:

  Frame the group's discussion into tangible issues

  Get the group organized to address the issues

  Work between sessions to maintain continuity and follow through.

- It is very important in follow-up workshops to bring and use copies of the original partnering documents.

- When internal organizational issues come up, it is helpful if members of the organizations can communicate openly with one another.

- It is easy to get lost in a short-term focus but essential in every workshop to take a block of time to focus on what's going to happen two, three, or more months into the future.

- It is very important to use partnering not only to solve current problems but also to identify and work on developing opportunities.

15
Partnering as Repair

The State Records Building Case

Partnering as Repair

"We never had a preconstruction workshop. Now the project is behind schedule, over budget. Is it too late now to use partnering, since we didn't start that way?"

In an ideal world, partnering always begins with a preconstruction workshop. In the real world, things happen.

Can partnering still be effective if it is used after a project has started? Is it possible to use partnering as "repair work" to get a project back on track?

Partnering looks different if it is used after a project has started, but the strategies and skills involved in partnering can still be useful. In fact, partnering can be very effective as a means to "repair" a project that has gotten off track. This case illustrates how "partnering as repair" can work.

Characters

Gary Braun. Integra Plumbing

Al Greenfield. Project Architect, Glass, Miles and Frank

Henry Gorman. Project Manager, Jordan Construction

Greg Johnston. Facilitator

Roy Jordan. President, Jordan Construction

Tom Keyes. Project Manager, Midwest Telecomm

Marion Little. Director, State Records Agency

George O'Malley. Project Engineer, Omni Structural

Ted Phillips. Principal-in-Charge, Glass, Miles and Frank

Mark Snyder. Project Leader, Barr Electrical

Fred Waters. Ryan HVAC

Bob Winters. Project Manager, State Records Agency

June 6: Glass, Miles and Frank Associates

"Have you lost your mind?" Al Greenfield bellowed over the filing cabinets. "We're two months behind schedule in the fourth month of construction, and you're asking me to take a whole day to attend some workshop?"

As Glass, Miles and Frank's project architect for the highly visible State Records Building project, Al was very worried about the mounting delays on the job. Every additional day of project work seemed to accomplish only half of what it was supposed to.

Walking around the cabinets into Al's office, Ted Phillips countered, "This should be different. It is one of those partnering workshops. We'd use a facilitator, somebody with experience. We've got to find some way to get ourselves back in front on this project."

As Principal-in-Charge of the Records Building project, Ted also worried about the delays. With 20 years experience, and much of it with government bureaucracies much like the Records Agency, Ted had lived through more than his share of bureaucratic messes. Nothing could match the current tangle, though, not even the big federal job he had worked on three years earlier.

"I'll try to keep an open mind," Al replied. "You want me to block out a complete day of my time, and yours too, at this critical point in the project, to meet with the same stubborn people we meet with every week already. Isn't the three hours we already spend with Jordan Construction and those folks from the agency every week enough?

"Everybody in town knows Jordan makes its money from claiming changes for all kinds of things that aren't necessary. And as far as the Records Agency goes, everybody knows that's just a haven of political appointments. Are you a glutton for punishment, meeting with these people more than we have to?"

"I keep telling you this is different," Ted went on. "This is partnering."

Al shook his head. "Don't tell me you've really fallen for that, and at your age. I've heard about that partnering from some of my pals over in the military agencies. They say it's just like the total quality fad, one of those things management buys into for a few months, then it goes away.

"What's new about partnering anyway? We've been partnering with our clients for years, haven't we? I think we even have it in the firm's mission statement.

Ted smiled, "You've become even more cynical than I thought. Come on, what's a day at this point? We don't have much to lose. You yourself keep saying that every day of work on the project puts us another half day behind. Maybe if we take a day off we won't fall behind as much.

"Besides, you know as well as I do that half of our government clients are using partnering. We might as well get some experience with it."

"Right," Al replied. "I give up, I'll go. I can't really do much for the project by working on it directly. Every time I put my foot down on that thing, I seem to slide backward. Go ahead, before I change my mind. Fax the agency our return slip there, mark the date selection for two weeks from Friday."

Ted marked the date Al selected. Maybe he would tell Al later that it wasn't entirely accurate to call this the Records Agency's workshop. After all, it was Ted himself who had suggested partnering to Marion Little, the agency's director.

Late May: Getting Partnering Organized

Ted had not gone directly to Marion Little with the idea for partnering without talking to Al. He had brought the subject up with Al a number of times, but Al kept deflecting him, arguing that the delays were being caused more by bad luck and a stubborn contractor than by communications tangles. Al also was against the idea of getting together with the project team; he didn't think anything could come of it.

Ted himself wasn't sure that anything could come of getting together. He had no experience with partnering, but he had read about it in several journals and discussed it at length with another project architect at the firm. The other project architect reported substantial, positive results in what Ted knew was a tricky client and subcontractor situation.

Because Marion Little had heard about partnering some months earlier, she was familiar with the concept. She didn't think it was possible to use it for the Records Agency's building because the bids and contracts had been finalized months earlier, and she didn't know how to make additions. She also had assumed partnering had to begin with a preconstruction workshop, and it was too late to organize something like that.

When Ted explained that, even without a preconstruction workshop, they could still use partnering for conflict resolution and to improve communications overall, Marion quickly agreed to try it. They agreed to try to schedule a workshop, and to attempt to share the workshop's costs with Jordan Construction. Roy Jordan, who had used partnering with positive results on a number of his firm's jobs, quickly consented.

Ted Phillips got a list of experienced partnering facilitators from the local office of the Associated General Contractors and narrowed the list to three candidates based on telephone conversations with five of them. The three who made his short list came in for a brief interview with Ted, Roy Jordan, and Marion Little. They settled quickly on Greg Johnston.

Johnston did not have the most experience, but his approach and expertise seemed best suited for this job. He had worked with a number of government agencies, and he approached partnering more as an integral part of the project than as an add-on seminar or training program. He was interested in the workshop, but even more interested in and enthusiastic about applying the results of the workshop back on the job site.

June 7: The Agency

"Great!" Bob Winters came striding into the office, waving a curled piece of fax paper in his hand. "The architects say they can make the partnering workshop. Looks like we're on."

So it really was going to happen, Marion reflected. She had never thought that the contractor or key subcontractors would waver on attending a partnering workshop. They all had attended other partnering workshops and knew the value of the process. Some of them had been asking Marion to schedule partnering weeks earlier.

However, Marion did have her doubts that Ted would be able to get Al to support the workshop. Al seemed so aloof and out of touch with the project. Plus, most of the project's problems seemed to begin with him. She could imagine it would be difficult for him to go along with a workshop where he might have to deal with the problems he is causing in the project.

"We should get started on the attendees' invitation note," she pointed out.

Bob nodded, "You're right. Johnston warned us it could be difficult to determine who should attend, but I was surprised when you and I couldn't wrap it up in an hour last week. Do you want to give it some time right now?"

"Not right now," Marion replied. "I have some other calls I have to make. Come back right after lunch, though, and we'll finish the list then."

"What list? Are you two planning something else to keep me overloaded?" Don Luccarello poked his head in from the corridor. Close to retirement and in a relatively unimportant position, Luccarello was a holdover, a survivor from the previous administration. Since Little's appointment, he walked a fine line between independence and insubordination. His main concern, Winters thought, was to keep making his light workload even lighter.

"Don't worry Don," Winters replied. "It doesn't affect you. We all know how busy you are." With what, Winters thought, no one really knew.

"Sorry I lost it," Winters commented to Little as Luccarello retreated, "but I had to say something. He gets under my skin. I don't believe I've seen him do an honest day's work since I've been here. Do we have to keep him?"

"We don't have to do anything," Little replied, "but he would be a tough one to let go. His connections run awfully deep. It wasn't a battle I wanted to fight, at least not yet.

"Anyway, don't worry about it for now. We've got some good things happening—that's where our attention ought to be going.

Like this partnering workshop. That should turn into another feather for our reforms."

"I suppose so," Bob agreed.

June 12: Jordan Construction

"Is this room O.K.?" Roy Jordan motioned toward the conference room, leading the small group of his staff and the partnering facilitator through the firm's corridors.

"This will be just fine," Greg Johnston acknowledged, taking note of the excellent skyline view out the window. "Have you got a flip chart we can keep our notes on?"

Roy spoke while Henry set up the chart, "We're glad you're here, Mr. Johnston. The Records Building job is a mess, we haven't seen anything like it in years. We're two months behind, and we're just four months into the schedule. The drawings they give us from Glass, Miles and Frank, they're just not up to the par we usually see. We don't like to talk about other firms, but I don't know about this." Roy paused, looked down.

"We want this job to work. We don't want to be difficult. The architects deal with us as if we're being difficult. I don't think we're difficult, we're just trying to get the job done.

"We've got plenty of men in the field, and they have lots of experience. It's true we've never built a project exactly like the one we're working on here, but we have built all kinds of other things, and we've worked with more than our share of government organizations.

"We've worked with partnering, too, Mr. Johnston, so we know what to expect. They had partnering on that bridge job we did last year, and we've got it on the two federal projects we're working on now in the district. We believe in it.

"The only concern I have about partnering here is that people don't use it to cover a performance problem, or a professionalism issue. I don't like to cover things up if there are real problems. Henry, now what about you?"

Henry Gorman looked at Greg Johnston, "We never like to name names, it's not a great way to do business. This architect, though, Greenfield. I don't know where they got him. They say he's got experience, that he's done some big jobs. I don't know, I never heard of the man, and I've been in this business awhile.

"He just doesn't get it. He doesn't give us what we need, no matter how clearly we ask. He doesn't give us what we do need, and then he gives us all kinds of other work that we don't need."

"Could you give me an example?" Greg Johnston asked.

"Sure," Henry replied, "lots of them. A month ago we asked a simple question. The drawings specify a special finish in the interior, we think we can make a substitution. A simple question. We've worked with the material they're specifying, it's hard to get these days. The quality of it that we've seen lately is questionable. I know when they wrote the specification it was probably all right, but this is a state job. It's old. They wrote the specification years ago, and things have changed. The manufacturer went out of business."

"And what did the architects have to say about this? Did they argue with you?" Johnston kept writing notes while he asked.

Henry smiled. "That's the thing, they don't argue. Everything is 'Fine, we'll look into it.' 'Fine, we'll check it out.' 'Fine, this.' 'Fine that.' I wish they would argue, it would at least give us something to work with. They're nothing but a bunch of politicians.

"Actually," Roy Jordan cut in, "it's not that Greenfield ignores the questions. He answers a number of them, and that's where he creates even more problems. The answers he gives don't give us any more information than we already had."

"Examples?" Greg Johnston asked.

"Let's put it in perspective," Jordan continued. "Four months into this job, we've got an RFI backlog that stretches back to the first month. I count 60 outstanding RFIs, many of them going back to the first and second month, and about a third of them are duplicates of initial questions, going back for more information on the same issue."

Johnston wrote some more notes.

"That's right," Henry continued. "The finish work is just the tip of the iceberg. There's a whole set of outstanding questions about the electrical, some more about the mechanical, and more about the plumbing. I've got people out there just sitting and waiting, while we wait for answers. Somebody's going to have to pay for this, you know."

Johnston kept writing. The list grew longer.

June 13: Greg Johnston's Interviews—The Agency

"Sure you can concentrate here?" Bob Winters asked, pulling a metal chair up to the meeting table in the trailer. Marion Little pulled her chair up next to Winters's. "It can get a little noisy at times, people are always coming and going."

"I'll be all right," Greg Johnston replied. "This way, I can meet with everyone I need to in one trip. Thanks for setting up the interviews."

"No problem," Winters responded. "We're eager to make this thing work. The Records Building project is a very big one for us, not in dollars so much as in visibility. Some people in this town would love us to fall on our face, just to show that things haven't changed since our new commissioner came on board.

"Things really have changed, though," Winters continued.

"Corruption and waste are gone, a lot of the people are gone. There are still a few of the holdouts, all right. But it got too uncomfortable for lots of them after Ms. Little came in. She's been a breath of fresh air."

Marion Little blushed, acknowledging the compliment.

"Sounds encouraging," commented Johnston. "So you want this project to go well."

"And it isn't," Little noted. "We're losing ground on the schedule every week, and we can't get it straightened out no matter what we try. We've tried longer meetings, more frequent meetings, conference calls, RFI logs, you name it. It seems the more we work at it, the deeper we get lost."

"Could you give me some specifics?" Johnston asked. "In the communications arena, what are some things that are going well, and some others that are not?"

"I don't like to name names, it doesn't fit with our style these days," Winters replied, uncomfortably. "But the architect just is not an action kind of guy. I think he thinks he's building a statement, not a building.

"Greenfield keeps talking about 'the integrity of the design.' First of all, that word *integrity* gets under a lot of peoples' skin around here. They think he's calling them liars. Second, I don't think anybody knows what it really means, including him.

"Finally and most annoying, there's no action to go along with his talk. He talks 'integrity of design,' yet he doesn't show a whole lot of integrity himself. He doesn't do the everyday work, like return phone calls, keep accurate meeting minutes, or get to appointments on time. In fact, there are some people on the project who don't trust him at all."

"I think you really said it best in the beginning," Marion Little added. "We need action here and Greenfield is an idea man, not an action kind of guy."

Greg Johnston made a few more notes in his book and thanked Little and Winters for their time. Then he called to the subcontractors who were waiting at the other end of the trailer.

Greg Johnston's Interviews— The Subcontractors

Mark Snyder, Barr Electrical; Tom Keyes, Midwest Telecommunications; George O'Malley, Omni Structural; Fred Waters, Ryan HVAC; Gary Braun, Integra Plumbing. Greg Johnston jotted the names of the subcontractors down in his notes, and then asked the same question he had asked of Marion Little and Bob Winters: "With respect to communications, what's working well at this point in the project, and what isn't?"

A few seconds of silence passed, then all four men spoke at once.

"It's a war zone out there," O'Malley described.

"Never seen anything like it," Waters elaborated.

"I can't see how they're going to get this one to work," Mark Snyder worried.

"The architect and the contractor are battling with each other, and we're caught in the middle," Tom Keyes added. "They're fighting about the RFIs, and we're the battle pieces. We'll be sacrificed in the end."

"Why is that?" Greg Johnston asked.

"Pure economics," Keyes replied. "Jordan Construction and Glass, Miles and Frank are large companies. They can stand to lose a few dollars on jobs. They'll make it up on the next large project they get.

"We're not like that, though. If there's money lost on this job for us, there's not much of a cushion." Keyes paused.

"Let me be more specific," he continued. "If there's money lost on this job for us, there's no cushion. Any one of us, our firms could be lost on this job. It's small potatoes for the big guys, but it's our life savings."

"What, specifically, is happening on the project that is threatening all this?" Greg Johnston asked.

"I don't like to name names," Keyes began to reply.

"Then I will," Mark Snyder cut in. "Somebody has to, before we all lose our shirts. They need to replace the architect. They say he's got a lot of experience, but I've seen nothing but screw-ups. I think they should be ashamed of the drawings they give us; they're a mess. Nobody could follow them."

"Oh, grow up," Keyes interrupted. "You see drawings like that on every job. These aren't so great, but they're no worse than what we see all the time anyway. I say the problem with the architect is that he doesn't understand how to manage all the communications on this job.

"Look, the architecture firm is sitting on about 50 RFIs. That's a lot of RFIs in anybody's book, but I don't even think it's the number that's the problem. I think that Greenfield has no idea how to manage them. I'm not saying the man is lazy or incompetent. He could be the best designer in the state for all I know. What I am saying is that he hasn't got a clue when it comes to managing information."

"What's your view of this?" Greg Johnston asked George O'Malley.

"You boys aren't going to like this," O'Malley replied, looking down. "I think the real problem is with Jordan's man on the job, Henry. I don't think it's the architect at all. I know that we all like to bash the architects and most of the time we're right. Henry's the problem here, though. He's filing RFIs for so many things that he would be testing any architect.

"Could you be a little more specific?" Greg Johnston asked.

"Sure," O'Malley responded. "I've seen a lot of the RFIs Henry files. He doesn't need to file them. Instead of going through all the paperwork of an RFI, he could just call Al Greenfield on the phone and ask him to answer the question. That way, he wouldn't have to wait for an answer at all.

"Henry's creating a lot of the problems with delays for RFIs by using the RFI mechanism for small points of clarification. It was never intended to be used that way. He's causing the delays himself by doing this."

"I don't know about that," Fred Waters rebutted.

"Neither do I," Mark Snyder added in.

"And neither do I," thought Greg Johnston.

"Are you smiling at all this, you think it's funny?" Snyder asked Johnston.

"No, not funny at all," Johnston replied, "but familiar. On these kinds of jobs, it always seems that both sides have their own story, and they both sound pretty good."

Greg Johnston finished the discussion with the subcontractors, then met with the architects and the engineers. The architects strongly corroborated George O'Malley's point of view that Henry Gorman was abusing the RFI procedure.

"On my last job, a private school dormitory, almost as big a job as this one, we did the whole thing with no RFIs at all. The problem here, and we all know it, is that Jordan is trying to make a few extra bucks by stretching the contract out, getting lots of add-ons," Al Greenfield contended.

The engineers didn't have much to say about the RFI backlog; that was not the part of the job that impacted them most. They were, however, very bothered by the way Al was running the weekly project meetings.

"It's not that I mind going to the meetings, I know it's part of my contractual obligation, and I know I can give and get some good information," Max Eng, the structural engineer, commented. "I just hate to see the meeting waste so much time.

"Mr. Greenfield is too easy-going. He lets some people ramble on and on and doesn't cut them off. At the same time, there are others of us who come to the meetings who never get enough time to present our information."

Greg noted the engineer's comments, thanked all of them for their time, and took a brief break before interviewing Sam Howard, Jordan's site superintendent.

The trailer was the only place where Greg could interview him, and there had been several interruptions. Still, Howard had given a good interview, pointing out a whole area of problem communications no one else had mentioned: confusion about jurisdictions. Because the state owned the building, the local inspection authorities were tripping and falling over themselves claiming responsibility for their respective fields.

As he sat and reviewed his notes at the end of the day, Greg thought the lines of discussion were drawn pretty clearly. There was a clear disagreement about the RFI issue, and both sides sounded right. The contractors were complaining that the RFI backlog was excessive, that a number of outstanding RFIs were taking much too long to get cleared up, and that overall, the architect was slow to answer to a point that it was delaying construction and costing everybody money.

The architects also seemed to be correct in saying that the contractors were overusing RFIs. Greg had logged numerous examples, from the subcontractors and the engineers, of RFIs that seemed unnecessary.

It would also be important to address the jurisdiction issues and the project meetings when the group met in the partnering workshop.

The Workshop: Concerns and Goals

"They just don't build them like this any more, do they?" Roy Jordan asked as he surveyed the agency's conference room.

"What would you call this, Mr. Designer?" Jordan asked Ted Phillips. "1970s government building high tech, with all the little lights, the small windows, the black floor?"

"Nineteen sixty-four actually, was when we designed it," Phillips replied. "It was built and completed in 1966."

"You boys were the designers? I'm sorry, I didn't mean to offend. Henry, I've put my foot in it and we haven't even started yet," Roy explained.

"No offense taken," Ted Phillips replied. "After all, at the time, the building won awards. If you can get past the fixtures, you'll actually see that the thing works pretty well. It still functions effectively, and we brought it in under cost and ahead of schedule."

"Lots of integrity in this design," Al Greenfield commented.

Greg Johnston listened to the conversations with only half of his attention. He focused on rearranging the seats into a circle, so people would be able to work together more effectively. As long as it had ample space, lights, access to coffee, and walls to post flip charts, the room didn't matter much to him at all.

Although he was impatient to have the group get started on the communications problems he had uncovered in the interviews, Greg Johnston knew he would have to have the group move slowly initially. He knew from experience that people needed to build a foundation before they could work together productively on the problems they faced.

Others must have been impatient, too. The whole group was present by 8:15 A.M., 15 minutes before the workshop's scheduled 8:30 starting time.

Greg began by briefly explaining the history and goals of partnering, reviewing the agenda for the day, and discussing his own experience and expectations for the day. Greg reminded the group that this was a workshop, not a lecture or a seminar, and that meant that they would be doing the work themselves to address the issues facing the project. To get started, he asked them to list their interests and concerns for the day. When he could see that most of the participants had listed several items, Greg broke the group of 20 into four groups of five people each.

He asked each group to appoint a leader to summarize its discussion to report back to the larger group. Then he circulated among the groups, listening to the conversations. After about 20 minutes, Greg asked the groups' leaders to report on the discussions. He made up a summary list of the hopes and concerns people expressed for the day (see Table 15.1).

The group's frankness surprised Greg and concerned him just a bit. Some of the people in the room had some serious reservations, and Greg wondered if those reservations would get in the way of their ability to work on the project's problems. At the same time, Greg was pleased that people spoke out. If they felt concerned, then it would be important that others know so they could respond to it.

In fact, Greg shared their concern. He, too, always worried that groups would not carry back to the job the promises and commitments they made in a partnering workshop.

Greg explained how the goals statement functioned. He noted that people usually did this at the very beginning of a project. Even though this project was four months into construction, he explained that he thought the time spent writing a goals statement would be well spent because the group still would be working together for another year and a half.

The group agreed, and set to work producing goals to guide the remainder of the project (see Table 15.2). As people settled into a new mix of subgroups to work on the goals, Greg was pleased to see that everyone was participating enthusiastically, with no holdouts or wallflowers.

Table 15.1   State Records Building Partnering: Interests and Concerns

Interests	Concerns
▪ Get down to business	▪ Follow through workshop results on the job
▪ Clear up the mess	▪ Good ideas, but no translation to action
▪ Get back on track	▪ Can of worms
▪ Understand each others' reasoning	▪ Talk for the sake of talk
▪ Resolve and clear up the RFI backlog	▪ Worry that people will hold back
▪ Devise systems that work continuously	▪ Not enough time
▪ Build trust and strengthen relationships	▪ These problems cannot be solved
▪ Make some progress	▪ Results will not equal personal costs of day

As members of the group came up to the flip chart to sign the completed goals statement, Greg posted a list of communications trouble areas on the project: RFI procedures, jurisdiction confusion, meetings, and access to the client.

Placing the cap back on the marker after he signed the goals statement, Al Greenfield sympathized with Greg, "Those are some pretty serious problems we're having in all those areas. We're all curious to see how you're going to solve this. What are you going to do?"

"Not a thing," Greg replied, continuing to write. "These are your problems, you're the only ones who can solve them. Which one would you gentlemen like to work on?"

Henry and Al stood there in silence.

"Good point," Henry acknowledged. "There really isn't anyone who can solve these things except us, is there?"

"Right," Al agreed. "So what partnering does is to get us together and organize us so that we can solve our own issues."

They walked back to their seats.

Greg turned to address the whole group. "Here's where this workshop will differ from one done before construction," he

Table 15.2 State Records Building Partnering: Goals
Statement

- Treat each other with respect
- Work to earn trust and build trust
- Follow through on your commitments
- Separate problems from people; disagreements are inevitable
- Aim for win–win situations; steer clear of winning at someone else's expense
- Talk directly to people if you have conflicts with them
- Make a reasonable profit
- Aim high, quality is free
- Recognize time is money, and respond to requests quickly
- Work as hard on communicating as you do at your technical tasks

pointed out. "Both kinds of workshops produce communications procedures, but the procedures we produce will have to begin by undoing a number of procedures that don't seem to be working. I'd like you each to choose one area you'd like to work on."

Greg explained to the group that, based on his interviews, there seemed to be four areas in which the group could develop communications procedures. Better procedures in these areas might not completely resolve all the project's problems, but they would make a significant difference in project communications overall.

Greg asked the group to feel free to suggest other areas in which they might develop procedures. No one suggested any additions. In fact, people were so eager to get to work on the list that Greg had to suggest several times that people take a break before they stopped their discussions.

The Workshop: Communications Procedures

Greg believed and had seen many times that groups have the ability to govern and lead themselves, if provided a proper forum, structure, and facilitation. He discovered that groups don't need a facilitator to provide answers.

In the few times Greg had been tempted to provide a group with recommendations to solve a problem, the solutions had all backfired. People were unable to implement the idea, or they lost interest after the workshop. It was when a group member apologized to Greg for being unable to implement "his" idea that Greg finally recognized that he needed to steer clear of the recommendations business.

What groups needed Greg for was not to provide the answers himself but instead to make sure that the groups' members tapped into their own insight and creativity. Working on communications procedures, design and construction groups were usually able to devise specific systems to respond to their own problems.

The RFI backlog in this case was a good example. From Greg's interviews, both sides seemed "right." The architect was very much behind in responding to the RFIs; the contractor seemed to be using RFIs too often. No immediate solution seemed obvious. Yet Greg was comfortable, based on numerous similar efforts, that the group would work something out.

The participants divided into groups to work on the four tasks. Greg asked for a balanced mix of people from different organizations to work on each task. Al and Henry did not have to be asked to work on the RFIs, they knew it was their main issue. Bob Winters joined along with George O'Malley and Max Eng.

Group Problem Solving

"Well, are we going to be reasonable about this, or was that goals statement a joke?" Henry Gorman led off, folding his arms.

"If you're going to approach it like that, we might as well not even start," Al Greenfield replied, pulling his chair a bit outside the circle.

"Gentlemen, gentlemen," George O'Malley interceded. "What about that goals statement we just worked on? Don't you know you've got to approach this in a spirit of compromise? You've both got be willing to bend a little."

Greg Johnston interrupted all the groups at that point, "Before you start working on your assignment, I want to point out a few things." The groups stopped their discussions.

"You're probably jumping right in to come up with ideas to solve the problem," Johnston continued. "Don't do it, you'll tie yourselves up in knots because you'll immediately come up with some reason why the idea won't work.

"Here's where we can take some ideas from all those companies that use total quality management task teams. Start by really looking at the causes of the problem. Too many groups get right to solutions without doing this. You'll usually do better if you discuss all the different factors that contribute to the problem.

"Since each of the groups here is made up of people from a number of different firms, you'll each probably bring a different understanding to how the problem impacts the job. It will be useful for each person to hear all these different perspectives in order for the group to come up with some actions that work across the board.

"Don't try to work on 'defining the problem': that's too one-dimensional. The problems we're addressing here are multidimensional, caused by lots of things. So try listing 'aspects of the problem.' And take your time. The more time you take describing the problem, the more likely it is that it will begin to solve itself.

"Once you've listed all the different aspects of the problem, and you should take at least a half hour to do this, you'll probably be eager to come up with the solution. Don't—it's still premature, and it will get you right back to either/or, or black/white kinds of thinking.

"Instead, list possible actions and at this point, don't rule anything out. Be a little wild here, the problem with most groups is that they go with the first idea they come up with.

"Only after you've got a list of at least a dozen possibilities should you designate just one or two things to actually try. The key word here is 'try,' don't forget. You're not going to solve whatever it is for once and for all, but you can come up with some things to try.

"Once you've got the things to try, then you can get specific. Who's going to try it? When? Where? Get some names, times, dates, and places down so we can know when things are going to happen.

"Oh, and use the flip chart pages to record everything. That way, everybody in your group can see everything at once and you'll have a nice visual when you present your ideas later.

"Questions?" Johnston scanned the room. "Very well, then, I'll circulate now and look forward to hearing from you all in about a half hour.

The RFI Backlog Problem: Breaking the Logjam

"Okay, let's get going," Henry Gorman began. "You're the architect, you do the flip chart." He gave Al Greenfield the marker. "Go ahead, list aspects of the problem."

"All right," Greenfield replied, heading the chart in clean, block letters. "How about if I start the list here with 'Henry's bossiness.' Why do you always try to control us?"

"That's good," Greg Johnston intervened from outside the group. "Write that down, 'Henry's bossiness.' Do you have a counterpoint, Henry?"

"Try 'Al takes things personally,'" Henry smiled.

Grinning in spite of himself, Al listed both points in clean, block letters.

"Now what are some other aspects of the problem?" Greg asked, moving to another group.

There was an awkward silence for a few seconds, then Al spoke, "Henry, do you get some kind of bonus for these things? Come on, be honest. Do they pay you based on the add-ons?

"I admit that we were slow on turning RFIs around initially but we've been making a real effort lately. Now it seems like even when we get something right back to you, it's still not right. It's like we're playing some kind of guessing game."

Henry replied, "We're pretty frustrated. We get your answers back but they don't tell us anything.

"As far as the money goes, no, it isn't like that. There are no bonuses for this. What there is, though, is big trouble if I lose money on the job. You see, for me, the RFI is a kind of loss prevention, a way of covering myself. The paper trail helps me cover myself just in case."

"List those as aspects of the problem," George O'Malley broke in. "One aspect is Henry's fiscal liability; the other is his need for documentation."

Al listed both.

"It's the documentation thing," he reflected, while printing on the chart. "Your need to document I really can't argue with, it's the way you do business. I should probably do more of it myself. It's just that doing it gets the whole project into a cover-yourself mentality."

"I know it does and I don't like it, but I don't know what to do about it," Henry replied. "Write it down, at least."

"Are you sure you don't like it, maybe just a little?" Bob Winters cut in, trying to phrase his thoughts carefully. "Sometimes, at least from my perspective, it does seem as though you're enjoying coming after us for all the details."

"Am I really that bad?" Henry reflected. "I like having something tangible in my hand, but I don't mean to be crazy about it, I really don't."

"Will you two stop?" George O'Malley pleaded. "Can't we list as a possible action that we all need to remember that RFIs are a fact of life? We don't have to read so much into them."

"I'm not questioning that you think you need all that you're asking for," Al continued. "I just wouldn't do it that way myself. I can't imagine what you're going to do with all that documentation after you get it."

"It's that word *documentation* I'm looking at," George O'Malley commented. "Who is the documentation really for, in the first place? Isn't it just your level of comfort, Henry?"

"That's the funny thing," Henry replied. "I don't even really look at the RFIs as being for me. I mostly look at them as documentation for the rest of the project. Maybe documentation is the wrong word after all.

"Maybe the better word here is *bulletin,* Henry went on. "I really look at each RFI as a news bulletin, a piece of information to get out to all the people on the job.

Henry continued, "You see, I don't really need the information for myself so much, but I have to always be on the lookout for uniformity and consistency on the job. If I'm trying to clear something up in an RFI, the answer is always going to go to 20 plumbers, or 30 carpenters. They all have to see the same thing.

"I can live with a fair amount of ambiguity myself. But when it comes to communicating information to everybody on the job site, that's where the consistency really comes into play. If there's the smallest possibility that the thing can be misinterpreted, it will be.

"At any given time, I'm dealing with anywhere between 30 and a 100 tradespeople: carpenters, plumbers, electricians, masons, you name it. They will read the same drawing a different way. That's why I'm such a stickler for detail."

"That's even worse than I thought." Al took up the conversation. "How can I possibly give you what you need, when you're the only one who knows it in the right kind of detail. You're the one who's in direct contact with all those people on the site.

"No wonder we've been feeling like we're in a guessing game with you about what kind of RFI will work best. You really are the only person who can judge that. What can I possibly do to make it right? No matter what I do, I'm destined to guess wrong about what you really need," Al concluded.

"Why try?" George O'Malley asked, with a hint of a smile.

"What I mean is, when he fills out an RFI, let him draft the response, too. Most of the time the man knows what he wants, or at least he has a pretty good idea. Instead of him putting you on the spot to come up with the right answer, let him draft a response and then you can either sign it or reject."

"Great," Al replied. "You're right, I like it. It won't solve all the problems, but it will solve a considerable number of them. Henry, what about you, is this O.K.?"

"I'd say it would be a big step. Will you really work with the RFIs if I write them?" Henry was just a bit skeptical.

"Will you really write them if I work with them?" Al countered.

"Let's give it a shot," they both concluded.

The RFI Backlog: The Underlying Problem

George O'Malley urged, "Come on, let's keep going. We've only got one action step here, I think there might be more.

Can we take a look at some of the specific RFIs in the backlog?"

"Sure," Al replied. "I came prepared with the log." He pulled an overstuffed binder from his briefcase and deposited it on the table in the center of the group.

"I might as well lead off with a confession," Al began, staring at the binder. "That log is long overdue. Before this workshop was scheduled, I was trying to keep track of the RFIs with yellow stickies."

The rest of the group looked at him, surprised.

"I know, it's not very sophisticated," Al explained. "But it always worked for me before. I never had to keep up with the volume of requests there is on this job.

"Guess I'd better write this down on the chart. Aspects of the problem: volume of RFIs and my inexperience with organizing an RFI log.

"So I made up a spreadsheet to log the RFIs and organized the binder. Of course, when I did, I found 11 RFIs that our office has been sitting on. I found the people in our office who were responsible for the answers and cleared up all 11 RFIs."

Al handed a stack of papers to Henry, "Here, I'll open with this, Henry. Take 11 RFIs off the top of the pile. I guess you'll want me to keep this up on a regular basis." Henry nodded as Al went on, "so I thought I would fold this log into our weekly meeting."

"Thanks," Henry replied. "You're a step ahead of me. Now I guess it's my turn."

He took the binder from Al, keeping his hand sealed over the edge. "OK, now I confess. I'll admit right from the beginning, there are a few of them that never should have been filed. We've got a few people asking questions who are just too compulsive, even for me."

"You really admit that?" Al was relieved. "I thought you would try to justify every single one of them. How many are you taking out?"

"Ten," Henry countered. "Al, you've got to help me out a little on this one. How can I keep my men on the job from using these things improperly?"

"Why don't you just manage them tougher?" asked George O'Malley. "List that as a possible action."

"Easy for you to say," replied Henry. "I have a hard time doing that when I can't respond to their questions myself. Also, I have to live with these people."

Bob Winters volunteered an answer. "Why don't you tell them you'll include an RFI review in your regular weekly project meetings? If people know ahead of time that their requests will be reviewed, publicly, they might be a little more careful about what they submit in the first place."

"What do you think, Henry, would it work?" Al was cautiously optimistic.

"Care?" Henry reflected, "Yes, they care a lot. I think the idea would work pretty well."

"That's two action steps," George O'Malley cut in. "Can we go for a third?" He took the RFI log book from Henry and started flipping through it. Thirty more stuck RFIs, give or take a few," he mused. I wonder where they are."

"In Bob's office, of course. I thought you all knew that," Al motioned toward Bob Winters. "Because we're building on state land, I have to clear a lot of the questions with the state."

"All of the RFIs?" Winters was incredulous. "Stuck in my office?"

"Not all 30, of course," Al continued. "Just 20. And why should this surprise you, Bob? After all, you're the one I give them to."

"Sure, and I pass them off to our buildings consultants, who have been telling me that they get them right back to you."

"I guess 'right back' is a matter of perception," Al replied, looking almost as surprised as Bob. "We count 20 RFIs stuck in your office, Bob," Al noted. "Check the book."

Looking confused and frustrated, Bob started flipping through the log book, "Why didn't you tell me? Why didn't you say something?" Bob wanted to know.

"I did," Al replied.

Bob was getting defensive, "My memory can't be that bad. Why don't I remember these? What about this one? Wiring change on the ground floor."

"Right," Henry cut in. "Our people wanted to know if they could substitute a better quality, less expensive wire for the specification. Also, they didn't think the wiring would fit in the conduit that was measured for specifications. With all the wires, telecommunications, and computer installation that's anticipated, the conduit will probably be too small."

"A reasonable RFI," Al commented, "so I passed it on to Bob, since the state has a definite policy about wiring and conduits."

"And I just talked to our consultant on that last week," Bob spoke slowly. "He was asking what had happened to his recommendations. He was concerned that his suggestions to go to a larger conduit might impact other aspects of the design."

"Never heard from the man," Al replied. "And I didn't want to be on the phone to you to do it. I just assumed you knew what was in the works, and where all the outstanding paper was. I didn't want to be a pest."

"I understand," Bob continued, subdued. "You wouldn't have heard from the consultant directly, don't forget. It's my assistant, Don Luccarello, who coordinates RFI correspondence. Luccarello, the holdover from the previous administration. Could he be the logjam? Let's look at the other outstanding RFIs."

They got through another four RFIs in the log book when Winters excused himself to go check on Luccarello.

"Don't do anything you'll regret later," O'Malley called after him.

"Now that is what I would call an action kind of guy," Henry Gorman commented, gesturing toward Winters.

"A what?" Al asked. "An 'action kind of guy'? That's a phrase I keep hearing on the job site. Can somebody tell me exactly what it means?"

Embarrassed at his blaming the RFI problems on Al, Henry replied, "I think you're actually a pretty good example of what it means yourself, Al."

"I'm just wondering how we log this on the flip chart, along with our other 'things to try.' So far, we've got Al's log book, we've got the idea that Henry will draft his own RFIs and we've got the idea of the weekly review of RFIs to keep the people writing them in line. What do we call this one?" See Table 15.3 for action steps in solving the RFI backlog.

Wrapping Up

"We'd like to hear from each group in a few minutes," Greg Johnston interrupted before anyone could answer O'Malley's question. Over his slight resistance, the group chose O'Malley to present its action steps. The larger group listened and applauded when he finished.

"Doesn't any one of you usually critical people have any critical comments, I mean, constructive criticism?" O'Malley asked.

Marion Little spoke, "I think we're all pleased with your ideas. The only thing I could say at this point is, 'Go for it.'" The rest of the group nodded in support.

The other smaller groups (working on jurisdiction confusion, improving group meetings, and access to the client) presented their recommendations with similar reactions from the larger group. Each small group had come up with realistic recommendations for its task.

Table 15.3   State Records Building Partnering: RFI Backlog Problem-Solving

Aspects of the problem	Possible actions	Things to try
▪ Henry's bossiness	▪ Read Henry's mind	▪ On the next ten RFIs, Henry drafts his own responses
▪ Al's taking things personally	▪ Henry manages subcontractors with more discipline on RFIs	▪ Open review of RFI log at weekly meetings
▪ Henry's fiscal liability	▪ Computerize all requests	▪ Al keeps RFI log and reviews it weekly with group
▪ Henry's (legitimate) need for documentation	▪ Remember RFIs are inevitable	▪ Bob manages agency's internal information flow more closely
▪ Documenting equated to covering your butt	▪ Homicide?	
▪ Need to use RFI as bulletin to large group		
▪ Some subcontractors use too many RFIs, for small issues		
▪ Volume of RFIs		
▪ Al's inexperience with RFIs		

The group working on jurisdiction confusion suggested that Bob Winters and Marion Little be informed more quickly and take a more active role in the inspection process. The group had tracked the problem back to confusion that existed among the inspectors about which of their agencies had what authority. The group wanted more input from the Records Agency since they had the best information about current rules and procedures. Winters and Little quickly agreed to cooperate and volunteered to meet with whichever inspectors wanted to tour the site.

The group working on improving meetings agreed that the problems that afflicted project meetings included the large amount of information that needed to get exchanged and the fact that exchanging the information involved very different kinds of work. The more simple show-and-tell, need-to-know updates that people needed from each other went well.

Meetings also typically tried to tackle real job problems that came up, and it was in this area that most of the meetings' shortcomings occurred: some people had strong interests and opinions; others were not concerned with particular problems. Participation tended to break down, noisy people dominated, and quiet people retreated. As a result, the meetings produced solutions that were less than optimal.

To remedy the problem, the group suggested that meetings be divided into two distinct phases. The early information exchange phase would remain much as it had been. In the latter problem-solving phase, though, the leader would clearly divide the group into smaller groups. That way people who wanted to work on a particular problem could do so. Also, several groups could work at once.

To increase the meetings' effectiveness further, the group suggested that Al share leadership with other people on a rotating basis. They reassured Al that this was not a criticism of him but simply an attempt to spread the "ownership" of the meeting among all group members.

Al enthusiastically accepted the recommendation, "Great idea. Now all of you can get some sense of how hard it is to control this group."

The final group had addressed the issue of improving access to the client. They explored the concerns many people on the project team had about Marion Little's bosses, the State Commissioners of Records. Appointed by the governor, the seven commissioners had a history of infighting. The project team worried that they might step in at any point, as they had on some previous projects, and reverse decisions Marion Little and Bob Winters had made. The project team wanted a clearer statement of the commissioners' goals and lines of authority.

Marion explained to the group that these perfectly reasonable questions had no answers. The seven commissioners had very different goals and their authority was unclear. There are some laws and rules but the group articulates its powers as it goes. It happens in the tone of office. It's like two different Presidents make the Presidency look like two different offices, except here there are seven commissioners, not just one.

Faced with this political reality, the group reformulated its task to at least get more timely input and involvement from the commissioners and to use Marion and Bob more effectively as connections to the commissioners. Marion told the group she could represent their interests to the commissioners more effectively if members of the project team would periodically attend the commissioners' monthly meetings. She also suggested that they learn to put all requests in writing. She knew that they did not enjoy writing, but pointed out that that was the way the commissioners did business.

The group also pressed Marion to develop faster ways to get at least some informal feedback from the commissioners in between the meetings. They pointed out that a month could be a long, and expensive, time period on the project if they were waiting for approvals and that it should be possible for Marion to poll the commissioners by phone informally.

Marion liked this idea because it gave her a reason to do something she wanted to do anyway: increase her contact with the commissioners. She agreed to develop an informal polling procedure with the commissioners' input and approval.

When all the groups finished their presentations, Greg Johnston asked them to reconvene as small groups and work at a more specific level on implementing their ideas: Who would carry out what tasks when, where? He asked the groups to list these actions on flip chart pages so they could be presented to the larger group and so they could be saved for the next meeting.

To conclude the workshop, Greg asked for written comments and for suggestions on when people thought would be an effective time for a follow-up meeting.

Commentary

The State Records Building case makes it clear that partnering strategies and skills can provide valuable benefits for project communications even after construction has begun. The case illustrates how some of the traditional components of partnering, such as the goals statement, can be used in much the same way in "repair" partnering as in comprehensive partnering.

The case also illustrates that other components of partnering, such as the resolution of specific conflicts, must be expanded. More specifically, the aspects of partnering that need to expanded for repair partnering are prework and diagnosis, problem solving, and communications procedures.

Prework and Diagnosis.   In the State Records Building case, Greg Johnston invests considerable time and effort into interviewing project team members before the workshop. This extensive prework is not a luxury but a necessity in using partnering with a project that has already begun because the project has a history.

It may seem obvious that ignoring the project's history in partnering would be ignoring the reality that project team members live with, and this would invalidate the value of any partnering effort. Yet we have heard of "canned" partnering workshops that have attempted to do this. We have even heard of partnering programs in which participants were asked to forget about the project while they were at the workshop in order to avoid being "too negative."

Even if participants are able to do this they inevitably comment later that "partnering isn't such a great thing after all; it didn't accomplish much for the project. It was kind of fun, but it didn't solve any of the problems we are facing."

The diagnosis is important for the facilitator, because it allows him or her to think about which problems the group should address when they work on communications procedures. Formulating these key areas of project issues effectively is crucial for the overall success of partnering. The group will work on what the facilitator assigns; the trick is to define and assign issues that will yield the greatest paybacks for the project.

The facilitator has a daunting responsibility. Ten, fifteen, maybe even twenty people will give up a day of their billable time to attend the partnering workshop. The facilitator has a responsibility to formulate issues for the group to work on in such a way that provides the greatest returns for the project.

The prework and diagnosis give the facilitator the time necessary to formulate the right issues. In theory it may be possible to do this formulating "on-line" in a workshop, but in real life it is difficult because the issues are complex. It would be easy to formulate issues to keep the group busy but not necessarily result in optimum yield for the project.

In addition, the time spent in diagnosing the problems helps the facilitator discern the issues more clearly. People in the design and construction business can frame and articulate their own concerns very clearly, it is easy to be swayed by someone who is articulate but biased. The State Records case illustrates how different people on the project team can define the exact same issue in completely opposing, different ways. Doing interviews and prework, the facilitator has the time to weigh these opinions and make independent judgments. In the dynamic environment of a workshop, it would be harder to discern the thread of truth in a web of bias and emotion.

Finally, the prework serves a function for the group as well as helping the facilitator. Sitting through an interview on project issues gets people on the project team thinking, and thinking in particular about their own role in contributing to communications problems. After an interview, people are much more likely to come to a workshop prepared to get to work on communications issues.

Problem Solving.   Projects that are even just a few weeks old have a history. To some, the history is an inconvenience because it involves inevitable problems and disagreements that may coast a "negative" tone over partnering. Although this may be true to some extent, these same problems also provide tangible opportunities to demonstrate that partnering works.

People at a preconstruction partnering workshop often express the sentiment, "This is all well and good but how do we know what's going to happen when we get back to the job?" People at a repair partnering workshop never have to address the "what if" question because they already know what has happened.

The problems that evolve on a job provide a major part of the curriculum for a partnering workshop. They constitute issues that must be resolved in order for the project to succeed. People who attend the workshop are usually preoccupied with the problems. They come to the workshop looking for results.

Because of the impact and immediacy of job problems, it is usually necessary to devote a major part of the workshop to addressing them and devising specific action plans.

Making problem solving an integral part of the workshop agenda is somewhat of an act of faith. People have to believe that it will be possible for the group to at least make some progress on a problem, if not to solve it completely. It is quite possible that some of the problems groups tackle will not be solvable. After all, the same people will be working on them in the workshop who have been living with them for weeks, even months. Why should the workshop be different?

Yet it has been our experience that groups are almost always able to make substantial progress on the issues they address. Partnering workshops provide forum, structure, and the skills necessary to make it happen. Because it is a public workshop, people work a little harder to be on good behavior.

Also, the workshop provides a facilitator who can train and mediate. We have had the experience may times of intervening in a group that has been stuck for some time on trying to solve a problem. We find that the simple three-step process described in the case helps unblock groups' logjams and speed discussion toward agreement.

Communications Procedures. It is still important, with repair partnering, to write a goals statement. Even though the project has begun, the goals statement provides the context for solving current project problems.

However, most of the repair partnering effort will focus on communications procedures. When groups get to work on solving project problems, their products will usually be improved procedures for project communications.

The State Records Building Case illustrates the types of communications procedures groups work on frequently. The RFI process gets most of the effort and attention because it is often both the most pressing problem and the issue that, if resolved, will contribute most to project communications overall. Improving meetings and client access are typical issues on any job.

Clarifying jurisdiction confusion is an important task on many jobs involving public sector clients because they are often caught in a web of ambiguous relationships with other agencies. Design and construction projects bring all the ambiguities to the surface and enable them to be played out in the project. Partnering can benefit the client agency significantly by enabling it to clear up organizational issues that it faces on numerous fronts.

Partnering with
Government Clients

The case illustrates four issues typically encountered in design and construction projects that have government agency clients:

- Ambiguity and confusion within the client organization, resulting in difficulties in turning paperwork around and providing information back to the project.

- Internal conflict resulting from past political appointments and "tenured" employees, resulting in the agency's difficulty in following through on policies.

- Confusion at the top of the organization stemming from a board or commission that does not usually act in accord or with predictability or consistency.

- Ambiguity and confusion among the client organization and other agencies, often played out in the building inspection process and in other approvals and permits.

Partnering cannot change these conditions but it does provide a forum in which many of the project problems stemming from these conditions can be addressed.

Partnering can also help agencies clear up internal ambiguities. Just as the agency in the State Records Building case exposes its own internal issues in the course of partnering, many government agencies find that partnering brings their internal issues to the surface.

Government organizations are not the only ones that react in this manner to partnering; partnering tends to surface the internal tangles of any organization that is involved in the process. Frequently, for example, architecture firms that participate in partnering discover that their internal processes for handling RFI turnaround are not working because the firm overall does not handle internal communications effectively.

The ability of partnering to provide a wedge into internal organizational problems can be a substantial benefit. We have actually worked with several government agencies that have used partnering intentionally to identify internal problems they were having a hard time addressing. Partnering provided external data and impetus for the agencies to address and resolve organizational issues they needed to address anyway.

A Construction Lawyer Looks at Partnering

Christopher L. Noble, Esq.

Hill & Barlow, A Professional Corporation
Boston, Massachusetts

As the construction industry is in a time of transition, so is the practice of construction law. Accordingly, there is not a single or monolithic "lawyer's perspective" on partnering and the related topics discussed in this book. There are, however, a number of constantly evolving "lawyers' perspectives," of which this is one.

The central elements of the partnering process (the goals statement, communications procedures, and conflict resolution process) have a familiar ring. Indeed, not long before the symbolic signing of the flipchart page, many of the workshop participants signed another document that embodied similar principles. It was called a contract.

I have become accustomed to hearing, "Well, now that the contract is signed, we can all put it in the drawer and forget about it."

The authors answer the question "What will my lawyer say about partnering?" in Chap. 3. However, Christopher Noble, Esq., addresses the same question with the invaluable insight of an experienced construction lawyer.

But until the advent of partnering I hadn't realized how many members of the design and construction team took this comment literally. Indeed, many of the participants in partnering workshops have no apparent memory of the goals, procedures, and processes that they just finished negotiating or bidding on. Their objective seems to be to reinvent, on a personal level, the very relationships that have recently been contractually established at the firm or institutional level.

I think I understand some of the reasons for this phenomenon, but I am uncomfortable with its implications. There seems to be an ominous discontinuity between the world as described in the construction contract documents and the world of the partnering sessions. The contractual world is replete with written notices, preconditions, deadlines, prescriptions, proscriptions, incentives, and consequences, all of which are intended to be *enforceable*. The concepts of compliance and breach are central to the culture of this world. The partnering world is, in a sense, countercultural. It rejects, or at least ignores, the value system represented by the contract documents and seeks to replace it with an alternative value system. The problem is that the contract documents cannot be made to disappear with the wave of a facilitator's wand. In the drawer or on the table, they are available to any party who wishes to enforce them in accordance with their terms.

This brings me to a second discontinuity: It is the one between the personalized, consensual conflict resolution procedures that emerge from the partnering process and the institutionalized, coercive procedures that will be imposed by the contract or the legal system as a "default position" should the partnering procedures fail to achieve a settlement. This discontinuity can result in quite a shock, especially to a former "partner" who has put the contract in the drawer and forgotten about its terms and conditions until he or she sees them recited in an arbitration demand or court complaint.

Many lawyers may see these discontinuities as a clash between the "new age" culture of partnering and the "real world" culture of contracts and courts. In my view, the truth is more complicated, and more significant. What we see here are fragmentary pieces of an increasingly deconstructed construction industry that is struggling to reconstitute itself on the eve of a new century. Owners, designers, and builders will have to work together to achieve this goal, and I think that lawyers and partnering facilitators can help.

By way of background, the American construction industry operated for many years primarily through a balanced and stable triangle consisting of the owner, the design professional, and the contractor. In this traditional model, the owner usually had a long-term interest in the building. In the 1970s and 1980s, the strength of the traditional owner was diluted through such trends as internal delegation from top executives to financial or facilities management departments, and the ascendancy of developers whose interest in the project ended with a postconstruction conveyance. Concurrently, the unitary responsibility of the contractor for construction of the building was often replaced by the fragmented duties of the construction manager, who owed the owner the process of construction but not the product of construction (separate parts of which were owed to the owner by trade—(née sub)—contractors). The design function was also often unbundled, with the owner procuring the services of different design disciplines from different sources.

These trends are continuing in the 1990s, with tighter margins and shorter schedules wreaking even more havoc on relationships among project participants. New teams form for each project, and now even the owner's role is being outsourced to program managers who arrive and depart with everyone else. Construction sites resemble singles bars, where strangers are making it up as they go along.

Countertrends have emerged in reaction to these developments. Some, like design/build, have grown out of a realignment of traditional industry participants. Others, like alternative dispute resolution, have resulted from the application to the construction industry of broader societal themes. Partnering, it seems to me, fits into this latter category. It marks the entry into the construction industry of organizational development and human resources professionals, and reflects their increasingly widespread influence in the theory and practice of American business.

Of all these trends and countertrends, partnering is the only one in which members of the construction bar have not been active and influential (notwithstanding the fact that the longest-running and most successful partnering show has been directed by the General Counsel of the U.S. Army Corps of Engineers). Lawyers have helped to create, design, structure, and manage a large number of significant construction industry developments, but we have been effectively demonized by partnering promoters and participants.

The absence of lawyers from the partnering table has undoubt-edly contributed to the discontinuities noted above. If we were there, perhaps we could help the partnering process become less insular and naive about the contractual, financial, and legal envi-ronment in which it is being carried out. Conversely, we lawyers might become less insular and naive about the constructive power of open communication and other partnering skills and values.

Although it is unlikely that lawyers will be routinely invited to partnering workshops, I hope that we can work with partnering facilitators to make a positive contribution to their curriculum. Also, in the long run, there may be opportunities for constructive collaboration between the legal and partnering communities. One such opportunity might be the potential development of a new interdisciplinary service to the construction industry, performed by an individual or firm that could be known as "project counsel." The services of project counsel for a particular construction project might include some or all of the following elements:

- At the earliest possible time, project counsel would work with the owner and any of the other project participants who may then be on the scene, to identify their goals, objectives, financial resources, management and technical capabilities, etc., on the basis of which project counsel would help to select a delivery system for the design and construction of the project.

- Project counsel would advise the owner on procedures for selecting other project participants that would be appropriate for and coordinated with the recommended project delivery system. Project counsel services would include developing selection criteria, drafting RFQs and/or RFPs, and administer-ing the selection process.

- On the basis of information developed during the initial phases of service, project counsel would produce an integrated, pro-ject-specific set of draft contracts for all of the key project rela-tionships. These draft contracts would embody the principles of consistency, fairness, and equity. They would incorporate pro-posed financial incentives, risk allocation, and the like, with the goal of maximizing the success of the project as a whole.

- Project counsel would participate in an interactive teambuilding process that would combine the current elements of contract negotiation and partnering. Using lawyers' mediation skills and the skills of experienced partnering facilitators, project counsel would help to guide the parties through procedures that would result in both an affirmation of their relationship and a set of enforceable agreements. This would take the place of the current bifurcated process, in which lawyer-assisted parties first go through an intensive, adversary contract negotiation, followed by a lawyerless partnering retreat in which the parties try to build productive "off-contract" working relationships.

- The contract documents for the project would incorporate a conflict resolution system that would apply to designers, constructors, and other project participants, and would be administered to the extent necessary by project counsel. As appropriate, the system would involve channels of open communication developed by the parties themselves, standardized claim documentation, step and/or facilitated negotiation, mediation, standing neutrals, binding arbitration, and the like. This is one of the most important ways in which the legal and facilitation skills of project counsel would be placed at the service of the project participants, since it would integrate their informal conflict resolution efforts with more formal ADR and legal procedures.

- Project counsel would work to harmonize the contractual and administrative structure of the project with insurance, bonding, and other risk management devices.

- Project counsel might have a particularly important role in the project closeout process, facilitating and documenting the efficient resolution of any open issues that have not been settled during design and construction. The goal would be to close out the project on a positive note and to avoid an end-of-project blowup.

- Project counsel would be available to address postconstruction disputes arising out of building failures. There could even be a role for project counsel in the resolution of third-party claims such as personal injury suits by individual plaintiffs, "sick-building syndrome" claims, etc. Project counsel's familiarity with the design and construction of the project, combined with a history of scrupulous neutrality, could be an important resource for the resolution of any disputes relating to the project.

The project counsel concept is being developed by the American College of Construction Lawyers, a group of experienced specialists who are exploring ways that the legal profession can contribute to the advancement of the construction industry. It is by no means the only way that lawyers and partnering facilitators could work together but it points to an important observation: Partnering will have a significant long-term impact on the construction industry only if it comes out of its workshop "closet," and becomes integrated with the other institutions that shape and define the design and construction process. Forward-looking construction lawyers are extremely interested in this potential and are eager to help in its realization.

Index

About the Authors

WILLIAM C. RONCO, PH.D. and JEAN S. RONCO, ED.M. are Co-Directors of the Center for Business Partnering at GATHERING PACE, INC., a management consulting firm located in Bedford, Massachusetts (617-275-2424). They have over 15 years' experience designing and implementing over 250 partnering, teambuilding, and training programs for clients, including Ameritech, ASC Services Corporation, AT&T, Berkshire Properties, the Boston Housing Authority, Boston University, the Export-Import Bank, the General Services Administration, Johnson Controls, LaSalle Partners, Lotus Development Corporation, the Massachusetts Institute of Technology, the National Association of Corporate Real Estate Executives (NACORE), the National Parks Service, the State of New York, the U.S. Postal Service, and the Veterans Administration.

WILLIAM C. RONCO was Professor, Northeastern University College of Business. He earned his B.A. at Rutgers University, his Ed.M. at Harvard University, and his Ph.D. in Urban Planning and Organizational Learning at the Massachussetts Institute of Technology.

JEAN S. RONCO was Vice President, Human Resources, Thomson and Thomson. She earned her B.A. at Boston College and her Ed.M. at Harvard University.